Rangeland Sustainability

Environmental and Societal Dimensions of Sustainable Development Goals

There is a global need for a structured examination of a number of environmental science projects that are producing innovative approaches to providing governments, communities, and people with outcomes from the specific themes of the United Nations Sustainable Development Goals (UN SDG) 2030 initiative. This book series takes a new and comprehensive look at interdisciplinary ways of applying environmental science, technology, and science communication to explain how environmental science serves the societal needs of specific communities around the world to achieve Sustainable Development Goals. Each of the books in this series directly explores the use of environmental science, technology, and associated approaches to environmental management in a range of biophysical, societal, and theoretical-scientific topics, specifically related to energy, science and other partnerships, food security, societal productivity and infrastructure, water resources, land resources, climate, ecology, and sustainable practices. This series provides a multi-volume resource for ecologists, engineers, natural resource planners, and professionals in other related disciplines at the local, regional, national, and global levels of decision making.

Series Editor: Ricardo D. Lopez

Climate Actions
Local Applications and Practical Solutions
Brenda Groskinsky

Rangeland Sustainability
Social, Ecological, and Economic Assessments
Kristie A. Maczko, John A. Tanaka, Aaron J. Harp, and Matthew C. Reeves

Rangeland Sustainability

Social, Ecological, and Economic Assessments

Kristie A. Maczko
John A. Tanaka
Aaron J. Harp
Matthew C. Reeves

CRC Press
Taylor & Francis Group
Boca Raton London New York

CRC Press is an imprint of the
Taylor & Francis Group, an **informa** business

First edition published 2022
by CRC Press
6000 Broken Sound Parkway NW, Suite 300, Boca Raton, FL 33487-2742

and by CRC Press
4 Park Square, Milton Park, Abingdon, Oxon, OX14 4RN

CRC Press is an imprint of Taylor & Francis Group, LLC

Library of Congress Cataloging-in-Publication Data
Names: Maczko, Kristie A., editor. | Tanaka, John A., 1955- editor. | Harp, Aaron J., editor. | Reeves, Matthew Clark, editor.
Title: Rangeland sustainability : social, ecological, and economic assessments / edited by Kristie Maczko, John Tanaka, Aaron Harp, Matthew Reeves.
Description: First edition. | Boca Raton, FL : CRC Press, 2022. | Includes bibliographical references and index. | Summary: "This book provides the only integrated description of the indicators of rangeland sustainability that captures ecological, economic, and social dimensions. It takes a fresh look at the information available on the current and emerging issues across rangelands, and presents collaborative research for future progress. Offering a framework for evaluating rangeland sustainability, the best data tools to use, as well as an interactive tool for assessing at a variety of geographical scales, readers with limited knowledge of rangelands, professional rangeland ecologists, and land managers, will understand the best tools available today to assess sustainability across the geography of rangeland in the U.S"-- Provided by publisher.
Identifiers: LCCN 2021055395 | ISBN 9780367482725 (hardback) | ISBN 9781032246796 (paperback) | ISBN 9781003279747 (ebook)
Subjects: LCSH: Rangelands. | Range management. | Range ecology.
Classification: LCC SF85.3 .R365 2022 | DDC 333.740973--dc23/eng/20220111
LC record available at https://lccn.loc.gov/2021055395

ISBN: 978-0-367-48272-5 (hbk)
ISBN: 978-1-032-24679-6 (pbk)
ISBN: 978-1-003-27974-7 (ebk)

DOI: 10.1201/9781003279747

Typeset in Palatino
by SPi Technologies India Pvt Ltd (Straive)

Contents

Series Preface

As humankind has moved fully into the 21st century, societies have shifted much of their infrastructural and scientific foci, by necessity, to not merely "environmental awareness" but additionally to the very immediate and practical interconnections between our environs and how those environs impact fundamental needs. The fundamental needs of humans cover the gamut, generally, from human safety and well-being, physical and other infrastructural elements, to the associated sociological components that define and shape our varied and complex communities around the world.

This book series and its various volume contributors seek to answer several practical questions that arise regarding sustainability, specifically focusing on the fundamental needs of humans, which are increasingly posing new challenges and opportunities every day to humanity. As humans ever-presently endeavor to address their fundamental needs in a changing environment, and as you read this book, you will notice a common theme, which is a long-standing desire by all of humanity to obtain a livable, comfortable, and, yes, perhaps even an enjoyable life for this generation and those of future generations – this fundamental desire and pursuit is generally referred to as human well-being. Humankind in its varied sociological forms around the world shares this fundamental value and desire for well-being in ways that are expressed diversely, in as varied of ways as the grains of sand among the many beaches of our planet.

From the institutional perspective on such topics as are addressed in this series, humankind has come together in a number of ways to find solutions to sustaining ourselves. One of these institutional structures that represent most of the many and varied societies of the planet is the United Nations (UN). In recent years, the UN has undertaken a deliberate focus on the key elements of human well-being in their established institutionally structured manner, by way of the UN Sustainable Development Goals (UN SDGs). Among the 17 articulated UN SDGs are the following 10 UN SDGs that focus on topics that tie directly to the environmental sciences and engineering, which is the primary focus of this series:

- Eliminating Hunger (UN SDG 2);
- Ensuring Clean Water and Sanitation (UN SDG 6);
- Developing Clean Energy (UN SDG 7);
- Stimulating Industries, Innovation, and Infrastructure (UN SDG 9);
- Supporting Sustainable Cities and Communities (UN SDG 11);

- Enabling Responsible Consumption and Production (UN SDG 12);
- Informing Climate Action (UN SDG 13);
- Protecting Life below Water (UN SDG 14);
- Fostering Life on Land (UN SDG 15); and
- Developing Partnership for Meeting Sustainable Development Goals (UN SDG 17).

This series seeks to explore each of the ten previously listed UN SDGs through the lens of ongoing and future work in the environmental sciences and engineering, integrating the importance of science partnership with specific communities and practitioners around the world.

The series volume you are currently exploring, ***Rangeland Sustainability***, is a rich summary, and current demonstration, of how the Sustainable Rangelands Roundtable has utilized science and inclusive engagement with communities and practitioners to address both **UN SDG 15 (Fostering Life on Land)** and **UN SDG 12 (Enabling Responsible Consumption and Production)**. This volume covers efforts across the United States that impact the effectiveness and the long-term viability of major land use worldwide, rangeland use and management, which not only impacts the livelihoods, food supply, and well-being of much of the world's population but also serves as an important driver of ecological conditions in rangeland areas and beyond. The exploration of sustainable use models and approaches in this volume holds hope for sustaining vast areas of land across the globe by balancing vital sociological and ecological factors. To this end, the editor and authors of this volume have laid out clear pathways to guide the reader in the variety of techniques, methods, and new perspectives, providing a truly valuable guide for scientists, practitioners, and communities.

This volume addresses issues of fostering life on the land, while enabling responsible consumption and production through a collaborative approach, which demonstrates for the reader the progress that can be made through effective partnerships to tackle challenging linked topics, such as climate change, energy development, food security, and ecosystem services. The forum, the Sustainable Rangelands Roundtable, and its progress are described by the volume editor and authors to demonstrate a number of collaborative approaches that are useful for scientists and practitioners. Accordingly, the forum's design and/or approaches are likely useful beyond the scope of this volume, too – that is, for a broad spectrum of readers who are facing problem-solving challenges in a number of other environmental science, and natural resource management, applications.

Ricardo D. Lopez, PhD *(Series Editor)*
October 2, 2021

Preface

Early advocates for sustainable development noted that sustainability is a journey rather than a destination, as goals may shift based on societal values and preferences. When the Sustainable Rangelands Roundtable (SRR) was founded in 2001, based on conversations dating back to 1999, partners elected to adopt the Brundtland definition as a frame for the evolution of criteria and indicators – *sustainable development…meets the needs of the present without compromising the ability of future generations to meet their own needs.* Throughout the subsequent 20+ years, sustainability moved from relative obscurity to a featured entry on pollster Frank Luntz's list of words to use and lose, appearing in the 'lose' column, and then back into the mainstream as agricultural supply chains adopted management philosophies emphasizing sustainable production of beef, pork, poultry, and other products.

In the same vein, varied rangeland assessment efforts date back to the 1930s. On April 28, 1936, the Secretary of Agriculture submitted to the Senate Committee on Agriculture and Forestry a report of more than 600 pages, completed by the US Department of Agriculture (USDA) Forest Service titled, *The Western Range – A Great but Neglected Natural Resource*, as a requisite response to Senate Resolution 289. This document was touted by the then Secretary of Agriculture as the first attempt to make an inclusive survey of rangeland resources, their original and then-present condition, causes and effects of changes, socioeconomic functions rendered to the West and to the nation, and recommended practical solutions for the more important problems. The report highlighted the relative obscurity of rangeland resources, as well as the wide scope of research and associated data needed to create a foundation for sound use and management, early recognition of adaptive management, and adjusted practices to address concerns reflected in research and monitoring data. Similarly, the Secretary of Agriculture identified a need for effective extension services to provide information derived from research data to landowners and managers and assist these stewards with implementation and necessary adjustments. The need for increased rangeland research, extension, and adaptive management persists today, the not-so-secret sauce required for rangeland sustainability and continued improvement, to address at-risk areas identified by assessing ecological, economic, and social indicators in this book.

Almost 40 years later, the Resource Planning Act of 1974 charged the USDA Forest Service with periodic assessment and reporting on the status of forest and rangeland resources, completed on a 10-year cycle, with 5-year updates and projections 50 years into the future. However, for rangelands, these reports have not been populated with standardized data at a national scale due to a lack of such information. While the USDA Forest Service Forest Inventory and Analysis program is charged with collecting ecological data on non-forested lands, funding has not been made available to do so. Similarly, the Soil and Water Conservation Act of 1977 gives the USDA Natural Resources Conservation Service comparable authority to work with related mission areas and report on resource conditions, to ensure that USDA programs for the conservation of soil, water, and related resources are responsive to the long-term needs of the nation. As reauthorized in the Food, Conservation, and Energy Act of 2008, reports and updates to appraisals are generally required on a five-year cycle. Despite these complementary pieces of legislation, a comprehensive report, covering both public and private lands, coast to coast, and border to border, has not been developed by the government; this book aspires to meet that need for rangeland resources.

In an attempt to offer a more expansive approach to ecosystem assessment, albeit without reporting on socioeconomic aspects, the H. John Heinz III Center for Science, Economics and the Environment initiated national environmental reporting with *The State of the Nation's Ecosystems: Measuring the Lands, Waters, and Living Resources of the United States*, released in 2002 and 2008. The first report examined 103 environmental indicators for six broad types of ecosystems (coasts and oceans, farmlands, forests, freshwaters, grass- and shrublands, and urban areas). The indicators in the report include soil biological condition, nitrate and phosphorus levels in streams and rivers, at-risk species counts in different types of habitat, plant and tree growth, non-native species trends, and carbon storage in various kinds of plant matter. The subsequent 2008 report included 108 indicators and highlighted challenges and opportunities, including increasing the availability of robust monitoring data; ensuring that indicators keep pace with scientific advances and changing policy needs; connecting national indicators across scales, with those used by states and local governments; and creating the institutional capacity to continue and expand monitoring and assessment into the future. Authors state that to cope successfully with the environmental challenges ahead, decision-makers and the public need timely, reliable, unbiased, scientifically rigorous information about changes and trends in the state of the nation's ecosystems, including rangelands, and call upon Congress to formally establish a set of national environmental indicators. These data needs are echoed with more specific references for rangeland systems within this book.

Healthy rangelands provide carbon storage, wildlife habitat, open space, recreation opportunities, forage for livestock, water filtration, and a host of additional ecosystem goods and services. Recent initiatives of the nonpartisan Western Governors Association highlight the intrinsic importance of rangeland systems and their sustainability. The latest effort, led by Idaho's Governor Brad Little, focuses on working lands and working communities throughout the Western United States, capturing critical linkages among ecological attributes of range and forest lands, and the socioeconomic systems of the communities that depend on these natural resources for quality of life and livelihood. Previous governors have highlighted challenges associated with forest and rangeland management. Tracking trends in their soils, water, plants, animals, productive capacities, socioeconomic aspects, and legal and institutional frameworks is crucial to measuring the success of such initiatives. The Western Governors' National Forest and Rangeland Management Initiative Special Report in 2017 recommended, specifically, integration of "rangeland assessment metrics (soil, water, plants, animals, productive capacity) to create a rangeland sustainability report that addresses ecological, economic and social impacts." Four years later, using 2017 data, we can follow the rangeland assessment document as a baseline to highlight examples of successful policy and management endeavors, assist in identifying areas where more concentrated efforts may be useful, and provide a foundation for informed discussion.

Similarly, at the national level, the Biden administration announced the Conserving and Restoring America the Beautiful initiative, a plan to conserve 30 percent of all US lands by 2030. The report notes that conservation efforts are more effective when rooted in the best available science and informed by subject matter experts calling for transparent and accessible information to increase shared understanding. However, with regard to complex rangeland ecosystems, it was also noted that no single metric or single database assesses progress, and as such, recommends that the US government create an American Conservation and Stewardship Atlas to collect baseline information on the amount and types of lands and waters managed for conservation. While extant national databases such as the US Department of Agriculture Natural Resources Conservation Service's National Resource Inventory and Forest Service's Forest Inventory and Analysis platforms are tasked with collecting rangeland data, methodological differences make integration difficult, and socioeconomic information collected by the USDA National Agricultural Statistical Service, US Department of Commerce Census Bureau, US Department of Labor Bureau of Labor Statistics, and many others is not mentioned. This book's contents complement the America the Beautiful initiative, providing a template for reporting not only ecological information, but also socioeconomic aspects of rangeland systems to evaluate sustainability and, ultimately, effects of rangeland conservation when overlaid with implementation data. Additionally, ecological, social, and economic assessment may guide adaptive management efforts, documenting outcomes of past and current management, and indicating the need for future adjustments. The America the Beautiful initiative calls on agencies to support the use of adaptive management to help adjust to challenges and opportunities, such as changing climate, shifting pressures, and new science.

In the private arena at more local levels, as the livestock industry that stewards private and public rangeland resources throughout the country creates a framework for assessing sustainable beef production, data analyzed and reported in this assessment may also aid their efforts. At present, the US Roundtable on Sustainable Beef (USRSB) is finalizing industry-wide goals for various segments of the supply chain, and the information presented herein will be relevant to high priority indicators that they've identified around land and water resources. Specifically for the cow–calf sector, the USRSB has included SRR's *Sustainable Ranch Management Assessment Guidebook* in their *Sustainability Assessment Guide* for producers. The 17 ranch-level indicators in the *SRR Guidebook* are derived from the national/regional scale indicators included in this book, and an electronic assessment tool has also been developed to aid producers in evaluating their operations (available at http://assessment.sustainablerangelands.org). Ultimately, this ranch-level tool facilitates a ground-up approach to social, ecological, and economic sustainability, complementing the regional/national data described in this book.

Sustainability has continued to evolve since the inception of the SRR, focused on a never-ending journey of continuous improvement, empowering us all to do better for our rangeland ecosystems. As the SRR approaches 25 years of open partnership efforts, more than a hundred people, including sociologists, economists, ecologists, policy experts, industry advocates, federal and state agency personnel, environmental groups, ranchers, and conservation professionals, have shared expertise and insights. This assessment is a reflection of their contributions, as much if not more so than the authors, and we thank them for their commitment, energy, time, and participation. It is our shared hope that the assessment information detailed in this book will instigate improvements at national, regional, state, local, and ranch scales; calibrate goals with the capacity of the land base; and design management systems, protocols, and policies to conserve rangeland systems, while simultaneously supporting the human communities that depend upon them.

Authors

Dr. **Kristie Maczko** is a Research Scientist at the University of Wyoming and Executive Director of the SRR. She was previously the codirector of Colorado State University's Rangeland Resources Institute. Dr. Maczko has spent over 20 years working with government agency personnel, policy experts, economists, sociologists, rangeland ecologists, professional societies, producer groups, and other stakeholders to advance rangeland sustainability. She works with interdisciplinary SRR research teams exploring rangeland ecosystem goods and services, soil health, rangeland sustainability assessment, and economic aspects of conservation practices and rangeland management for livestock and wildlife. She serves on the boards of the US Roundtable for Sustainable Beef and the Plank Stewardship Initiative, as well as the National Grazing Lands Coalition's Advisory Committee. She earned her PhD in ecology from Colorado State University, MS in recreation from the University of Memphis, and BA in journalism and BS in natural resource management from Rutgers University.

Dr. **John Tanaka** is currently an Emeritus Professor at the University of Wyoming and works part time for the SRR. He has served in numerous roles in the SRR since its inception including several years as Director. He has a BS and MS in rangeland resources from Oregon State University (OSU) and a PhD in range science with an emphasis on range economics from Utah State University. He was a rangeland economist with Oregon State University and a research scientist at the OSU Eastern Oregon Agricultural Research Center, focusing on sustainable ranching.

Dr. **Aaron Harp** is currently working part time for the SRR, focusing on social indicators of sustainability. He earned a BA in economics at California State University, Sacramento; an MS in agricultural economics; and a PhD in rural sociology, both from Penn State University. He was an Assistant and then Associate Professor of Rural Sociology in the Department of Agricultural Economics and Rural Sociology at the University of Idaho from 1992 to 2000. Since 2000, he has consulted for government agencies and private entities seeking insights into relationships between natural resource and public land policies and the economic and social organization of rural communities.

Dr. **Matthew Clark Reeves** (Matt Reeves) is a Research Ecologist with the Rocky Mountain Research Station in Missoula. He received his BS in range management (1995), MS in environmental resources (1999), and his PhD in remote sensing and ecological modeling of rangeland and agricultural environments (2004). Matt specializes in applying remote sensing and modeling to characterizing ecological dynamics of rangelands. His research follows four basic themes, including climate change, vegetation dynamics, anthropogenic disturbance, and decision support tools, such as digital state and transition simulation models, for improving management efficacy.

Acknowledgments

Foundations for materials presented on the following pages were established in the late 1990s when Dr. Larry Bryant and Dr. John Mitchell initiated the first stakeholder discussions to gauge interest and support for a roundtable process to identify criteria and indicators for assessment of ecological, social, and economic sustainability of rangeland systems. Outcomes of these preliminary conversations fostered strong engagement among diverse groups to develop a framework for consistent, periodic rangeland monitoring and assessment, coast to coast and border to border. Dr. John Mitchell and Dr. E. Tom Bartlett came together as co-conveners to formally create the SRR in 2001. Following Dr. Bartlett's retirement, Dr. R. Dennis Child assumed the role of co-convener with Dr. Mitchell, serving in this role until he passed away. The shared vision and dedication of these leaders made possible identification of 5 criteria and 64 indicators of social, ecological, and economic rangeland sustainability, as well as the conceptual framework for evaluating indicator interactions and integration. Their contributions created opportunities and enthusiasm for the assessment presented herein, as well as related materials developed by SRR during subsequent decades. Their efforts comprised critical first steps along this journey of continuous improvement for rangeland systems, science, and research; we are forever grateful for their enterprise and achievements in expanding rangeland sustainability.

Authors would also like to acknowledge the contributions of Michael Krebs, who provided invaluable assistance creating maps contained in subequent chapters.

A list of abbreviations used in this book is provided in Appendix.

1

Introduction to the Sustainable Rangelands Roundtable and the National Assessment

Extension of sustainable development principles to land management gave rise to sustainable resource management as a logical conceptual expansion (Shields and Bartlett, 2002). Operationalization of this concept required progress in several areas, including the development of indicators for specific sectors, such as forestry and rangelands (Billharz and Moldan, 1997). As a result, in the latter part of the 20th century, the United States opted to approach identification of resource-specific criteria and indicators for quantification of social, economic, and environmental factors through a series of stakeholder roundtables focusing individually on forests, rangelands, minerals and energy, and water resources. These discussions led to the proposal of a Sustainable Rangelands Roundtable (SRR)[1] in 1999. Formal initiation of the roundtable process for criteria and indicators of rangeland sustainability began in 2001.

To date, the most widely accepted definition of sustainable development was put forth in the Brundtland Commission Report, "Our Common Future," which stated that development is sustainable when it effectively "meets the needs of the present without compromising the ability of future generations to meet their own needs" (WCED, 1987). Although the Millennium Ecosystem Assessment references "sustainable" or "unsustainable" almost 100 times throughout the text, an operational definition of the term is not offered (MEA, 2005). Following publication of the Brundtland Commission Report, sustainable development continued to gain international prominence and acceptance, providing the theme for the World Summit on Sustainable Development (WSSD, 2002), held in Johannesburg, South Africa, during September 2002. A decade earlier, sustainable development also served as the organizing theme for the 1992 United Nations (UN) Conference on Environment and Development at which many countries adopted the Statement of Forest Principles and Agenda 21 as the basis for efforts to implement the concept (UNCED, 1992a).

The Statement of Forest Principles built upon the spirit of the Brundtland definition, recognizing its applicability to sustainable forest management and asserting that "forest resources and forest lands should be managed sustainably to meet the social, economic, ecological, cultural and spiritual needs of present and future generations" (UNCED, 1992a). Agenda 21 advocated

DOI: 10.1201/9781003279747-1

implementation through the creation of indicators of sustainable development as a basis for decision-making (UNCED, 1992b). As a result, in 1994, Canada organized an international Working Group on Criteria and Indicators for the Conservation and Sustainable Management of Temperate and Boreal Forests, known as the Montreal Process.

The Montreal Process produced a consensus list of 7 criteria and 67 indicators for sustainable forest management, ratified by the 1995 Santiago Declaration (Coulombe, 1995). Significantly, the 12 countries and 5 continents involved in the Montreal Process encompass 60% of the world's forests (USDA Forest Service, 2002) and approximately 75% of the world's rangelands (Graetz, 1994).

Within three years, the United States Department of Agriculture (USDA) Forest Service, which manages over 180 million acres of forests and rangelands, committed to the implementation of the Montreal Process criteria and indicators on National Forest System lands by 2003 (Dombeck, 1998). The use of performance indicators within a land management organization may be described as part of an adaptive cycle, with periodic monitoring and assessment documenting outcomes and suggesting areas for improvement, as shown in Figure 1.1.

Monitoring is a key source of information used in measuring performance in management organizations. Periodic assessment creates trend data to indicate the trajectory of change. Performance indicators provide accountability for decisions and actions, but they also promote the evolution of management practices that are more effective at achieving the goals and objectives of the organization. They provide the feedback that allows reviews

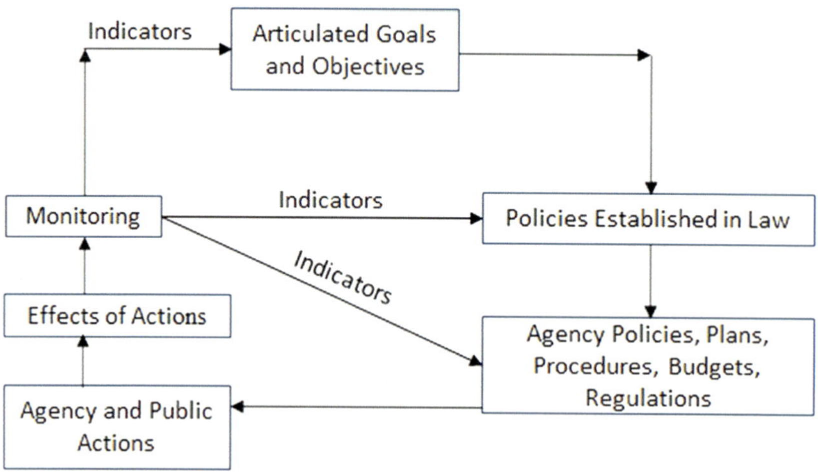

FIGURE 1.1
Use of indicators in United States' policy development (Heintz, 2002).

by on-the-ground managers, mid-level managers, and policy-level decision-makers from ranchers to county commissioners, and agency personnel to state legislatures and congressional staffs. At each stage, indicators can be used to determine successes and failures from the observed consequences. Information feedbacks help shape the evolution of policies and management practices to better achieve goals and objectives. Adaptive management processes allow organizations to repeat effective actions and discontinue efforts that fail (Heintz, 2002).

To facilitate implementation of the Montreal Process criteria and indicators, the USDA Forest Service convened the Roundtable on Sustainable Forests (RSF) in 1999. Due to the broad range of resources managed by the USDA Forest Service, at its first meeting, the RSF identified the need for development of a parallel effort to focus on criteria and indicators for rangelands. This idea warranted additional attention, and a scoping meeting was organized to investigate the utility of a roundtable for sustainable rangeland management. Efforts of USDA Forest Service personnel subsequently led to the formation of the SRR in April 2001.

People and Process

The SRR continues to operate 20 years later as an open, inclusive partnership with representatives from conservation groups, livestock industry, various levels of government, and academic partners committed to the task of using collaboratively identified attributes for assessing progress toward sustainable management of US rangelands (SRR, 2002). The SRR criteria and indicators published by the SRR in 2010 (Mitchell, 2010) described specific elements that combine to illuminate the path to long-term sustainability through elucidation of trends in environmental conditions, management actions, economic benefits, and social values provided by rangeland resources. SRR products offer a flexible framework for standardized data collection and reporting on the status of US rangelands to enhance national resource prioritization and planning efforts.

Since 2001, more than 100 organizations (Table 1.1) and over 150 individuals have joined in the SRR's mission to identify indicators of sustainability, based on social, economic, and ecological factors, to provide a framework for national assessments of rangelands and rangeland use. The assessment herein represents the first attempt to fully populate as many indicators as possible with scale-appropriate data and illustrate status of rangeland resources and the communities that depend upon them at a point in time. Neither the SRR nor the assessment authors attempt to make or advocate policy decisions; rather, we seek to provide opportunities for stakeholders from a wide

TABLE 1.1

Organizations That Have Participated in the Sustainable Rangelands Roundtable

- Agricultural Research Service
- American Farm Bureau
- American Farmland Trust
- American Sheep Industry
- Arizona Beef Council
- Arizona Cattle Grower's Association
- Arizona State University
- Boulder County Parks & Open Space
- Bradley University
- Bureau of Indian Affairs
- Bureau of Land Management
- California Rangelands Conservation Coalition
- Chippewa Cree Tribe
- Colorado State University
- Confederated Tribes of Warm Springs
- Consortium for Science, Policy & Outcomes
- Council on Environmental Quality
- Eastern Oregon Agricultural Research Center
- Ecological Society of America
- Economic Research Service
- Environmental Protection Agency
- Florida Fish and Wildlife Conservation Coalition
- Forest Service
- Foundation for Food and Agricultural Research
- Gray Ranch & Malpai Borderlands Group
- The H. John Heinz III Center for Science, Economics, and the Environment
- Hopi Tribe
- Idaho Conservation League
- Idaho State Department of Lands
- Idaho State Department of Agriculture
- Idaho Rangeland Conservation Partnership
- Invasive Species Advisory Committee
- Lady Bird Johnson Wildflower Center
- Mendocino Land Trust
- Montana Department of Natural Resources and Conservation
- Montana State University
- National Academies of Science
- National Agricultural Statistics Service
- National Association of Conservation Districts
- National Association of Counties
- National Association of State Foresters
- National Cattlemen's Beef Association
- National Grazing Lands Coalition
- National Institute for Food and Agriculture
- National Park Service
- National Wildlife Federation
- Natural Resources Conservation Service
- Natureserve
- New Mexico State University
- Noble Research Institute
- Northern Arizona University
- Oak Ridge National Laboratory
- Oklahoma State University
- Oregon Cattlemen's Association
- Oregon State University
- Pacific Northwest National Laboratory
- Plank Stewardship Initiative
- Point Blue Conservation Science
- Public Lands Council
- Quivira Coalition
- Ridley Block Operations
- Rocky Mountain Elk Foundation
- San Antonio Water System
- se4 Consulting, Inc.
- Society for Range Management
- Soil Health Institute
- South Dakota State University
- Sun Ranch Institute
- Texas A&M University
- The Nature Conservancy
- The Sonoran Institute
- The Wilderness Society
- Tribal Advisory Council
- Triple X Ranch
- University of Arizona
- University of California
- University of Colorado
- University of Idaho
- University of Nevada, Reno
- University of Wyoming
- Utah State University
- US Department of Agriculture
- US Department of Defense
- US Department of Defense – Avon Park Air Force Range
- US Department of Energy – Idaho National Laboratory
- US Department of Energy – Pacific Northwest National Laboratory
- US Department of Energy – Great Plains Research Laboratory
- US Department of the Interior
- US Fish and Wildlife Service
- US Geological Survey
- US Government Accountability Office
- US National Guard
- US Roundtable for Sustainable Beef
- Utah State University
- Washington State University
- Western State Land Commissioner's Association
- Wildlife Management Institute
- World Wildlife Fund
- Wyoming Business Council
- Wyoming Wildlife and Natural Resources Trust
- Wyoming State Grazing Board
- Wyoming Stockgrower's Association

range of backgrounds, representing potentially conflicting interests, to discuss information and perspectives on appropriate criteria and indicators of rangeland sustainability. Moreover, we identify research needs and discuss underlying data issues that need to be addressed before subsequent assessments are conducted. While interpretations or conclusions derived from the criteria and indicators may be contentious, the inclusive, open nature of the roundtable process ensures that the criteria and indicators developed by the SRR provide common ground for ongoing discussions about rangeland sustainability.

SRR organizers emphasized fostering trust and common understanding among participants. Associated rules and standards that guide roundtable practices and procedures are enumerated in guiding principles (Table 1.2) collectively created by SRR participants. Mission and vision statements originally crafted by SRR participants in 2001 embody this broad-minded approach to sustainable rangelands too. With regard to outcomes of the SRR process, participants agreed to a mission to move toward a future in which we have widely accepted criteria and indicators for monitoring and assessing the economic, social, and ecological sustainability of rangelands. Projecting more general expectations for rangelands, SRR participants seek a future in which rangelands in the United States provide a desired mix of economic, ecological, and social benefits to current and future generations. As the roundtable evolved to address data needs for emergent issues such as climate change, energy development, food security, and ecosystem services, the mission of the organization also adjusted to promote social, ecological, and economic sustainability of rangelands through conducting research, developing resources to communicate findings, and providing a forum for networking and collaboration.

TABLE 1.2

Guiding Principles of the Sustainable Rangelands Roundtable

1. Collectively, indicators should assess rangeland sustainable management at a national level; however, whenever possible, indicators should also be applicable at local levels.
2. Indicators should employ the appropriate temporal and spatial scales for assessing criteria and indicators.
3. Collectively, the indicators will address all three aspects of sustainability – ecological, economic, and social.
4. The Montreal Process Criteria and Indicators were considered as an initial framework; however, we also reviewed and considered other indicator initiatives.
5. There are numerous policy questions relating to rangelands. We have focused on the mission and vision statements agreed to by the SRR.
6. The roundtable process features outreach to stakeholders, open dialogue, and respect for differing opinions.
7. The SRR supports and strives to be compatible with improved on-the-ground management of rangelands.

While financial support for the SRR has been provided through grants and agreements with numerous organizations—including the USDA Forest Service, USDA Natural Resources Conservation Service, USDA Agricultural Research Service, USDA National Institute of Food and Agriculture, Western Sustainable Agriculture Research and Education program, USDI Bureau of Land Management, and US Geological Survey—the most significant support through the past 20 years has come from dedicated participants who have devoted considerable time and effort to the process. In addition to attending meetings and workshops, participants also assist with outreach and education efforts, perform field projects, voluntarily serve on special working groups, and respond to electronic queries distributed as part of a collaborative Delphi process. The critical contributions and commitment of SRR participants to the progress and success of criteria and indicator development for sustainable rangelands cannot be overstated.

Through 14 meetings held over a 26-month period from spring 2001 through summer 2003, SRR participants moved forward from Montreal Process work and crafted a suite of 5 criteria and 61 indicators to assess rangeland sustainability in the United States. The criteria are statements of categorical conditions or processes that are goals of sustainable development. Indicators are measurable, repeated observations of natural and social phenomena that represent systematic feedback, providing quantitative measures of human well-being, the economy, and natural resource conditions to illustrate trends and changes (NRCBSD, 1999).

SRR's work benefited from professional facilitation, which ensured that all voices had an opportunity to be heard. Diverse stakeholders began by considering the work of previous resource sustainability efforts related to forest management, and more broadly, the overall state of the nation's ecosystems. Through a series of discussions, participants identified suites of issues confronting ecological, social, and economic sustainability of rangeland systems in the United States. These spanned the spectrum from on-the-ground landowner concerns to department-level regional and national management priorities. Issues were then grouped into more specific, overarching areas of interest, which eventually gave rise to the definition of five broad categories or criteria. The five criteria are as follows (Bartlett et al., 2003):

1. Conservation and maintenance of soil and water resources on rangelands.
2. Conservation and maintenance of plant and animal resources on rangelands.
3. Maintenance of productive capacity on rangeland ecosystems.
4. Maintenance and enhancement of multiple economic and social benefits to current and future generations.

5. Legal, institutional, and economic framework for rangeland conservation sustainable management.

SRR participants developed measurable indicators under each criterion through extensive expert discussion and consideration of data sources. The absence of a data source did not exclude an indicator from inclusion, but its aspirational status was noted. While SRR participants used Montreal Process indicators established by the RSF as a source of information, other external sources were also consulted and new materials created. Programs extant in the early 2000s as SRR indicators were developed include the Natural Resources Conservation Service's National Resource Inventory (NRI, only nonfederal lands) and the USDA Forest Service's Forest Inventory and Analysis (FIA) platforms (although plot density on rangelands is sparse or nonexistent in many cases) for ecological data collection; the Forest and Rangeland Renewable Resource Planning Act of 1974 (RPA) that requires decadal reporting from the Forest Service (e.g., Reeves and Mitchell, 2012) for which criteria and indicators were used in 2000; a system of qualitative rangeland health measures (Pellant et al., 2005); the Environmental Protection Agency's Environmental Monitoring and Assessment Program (EMAP) (De Soyza et al., 2000), the USDI National Park Service (NPS) Vital Signs Monitoring (Fancy et al., 2009); the H. John Heinz III Center for Science, Economics, and the Environment's state of the nation's ecosystems report (Heinz, 2002); and initial attempts by the Nature Conservancy to explore ecoregion-based conservation (TNC, 1996).

SRR used a six-point framework to structure and evaluate their work (Mitchell, 2010):

1. What the indicator is, and what it measures.
2. Importance of the indicator as a measure of the criterion. This section included the indicator's scientific importance, based on the literature, and its robustness to changes in technology, as described by the NRC (2000).
3. Geographic applicability of the indicator throughout the United States.
4. Spatial and temporal scales of the indicator. An indicator valid at multiple scales can be more useful.
5. Data availability according to the following categories:
    a. Methods and data exist to monitor the indicator on all rangelands.
    b. Methods exist to monitor the indicator, but usable data are not available.
    c. Data sets only partially exist and/or methods are not standardized for the indicator.

   d. Although the indicator is conceptually feasible, no methods or data exist to monitor it.
6. The degree of understanding that stakeholders and the general public have for the indicator.

In addition to in-person meetings of topic experts, the SRR used a modified Delphi process to allow SRR members to continue discussion and decision-making in the interim. This process also maintained the engagement of participants who were unable to attend all meetings, eliciting their input on salient issues and decisions. The Delphi process facilitates systematic gathering and integration of expert opinion to reach collaboratively agreed-upon conclusions (Linestone and Turoff, 1975). Historically, it has been employed when data-driven methods are impractical. This method has enhanced policy decision-making, resolved environmental disputes, and assisted with economic planning (Miller and Cuff, 1986).

The iterative Delphi process allows experts to anonymously respond to questions. Individual answers are aggregated and sent back to participants, along with summary analyses and comments. Participants can revise their original answers in response to group feedback until a predetermined level of consensus is achieved. The SRR approach was identified as a "collaborative Delphi" because experts were not selected solely to respond to Delphi questions but rather for participation in the overall SRR process (Rowe, 2002).

SRR Criteria and Indicators

Criteria and indicators (Table 1.3) developed by the SRR describe individual elements to be monitored to determine trends in resource conditions, management, economic benefits, and social values derived from rangelands. Improved efficiencies that land management agencies and stakeholders can derive from this process include the following:

- directing monitoring and assessment efforts to areas identified by indicators as being important;
- instigating development of common data collection techniques and implementation of standardized, periodic inventory, monitoring, and reporting on public and private rangelands;
- focusing research by agencies, universities, and organizations on developing methods to measure and interpret criteria and indicators, and identifying research gaps;

TABLE 1.3

Criteria and Indicators for Sustainable Rangeland Management

Criterion 1. Conservation and Maintenance of Soil and Water Resources on Rangelands

Soil-Based Indicators:

1. Area and percent of rangeland soils with significantly diminished organic matter and/or high carbon:nitrogen ratio
2. Extent of rangelands with changes in soil aggregate stability
3. Assessment of microbial activity in rangeland soils
4. Area and percent of rangeland with significant change in extent of bare ground
5. Area and percent of rangeland with accelerated soil erosion by water or wind

Water-Based Indicators:

6. Percent of water bodies in rangeland areas with significant changes in natural biotic assemblage composition
7. Percent of surface water on rangeland areas with significant deterioration of the chemical, physical, and biological properties from acceptable levels
8. Changes in groundwater systems
9. Changes in the frequency and duration of surface no-flow periods in rangeland streams
10. Percent of stream length in rangeland catchments in which stream channel geometry significantly deviates from the natural channel geometry

Criterion 2. Conservation and Maintenance of Plant and Animal Resources on Rangelands

11. Extent of land area in rangeland
12. Rangeland area by plant community
13. Number and extent of wetlands
14. Fragmentation of rangeland and rangeland plant communities
15. Density of roads and human structures
16. Integrity of natural fire regimes on rangeland
17. Extent and condition of riparian systems
18. Area of infestation and presence/absence of invasive and other non-native plant species of concern
19. Number and distribution of species and communities of concern
20. Population status and geographic range of rangeland-dependent species

Criterion 3. Maintenance of Productive Capacity on Rangeland Ecosystems

21. Rangeland aboveground biomass
22. Rangeland annual productivity
23. Percent of available rangeland grazed by livestock
24. Number of domestic livestock on rangeland
25. Presence and density of wildlife functional groups on rangeland
26. Annual removal of native hay and non-forage plant materials, landscaping materials, edible and medicinal plans, and wood products

Criterion 4. Maintenance and Enhancement of Multiple Economic and Social Benefits to Current and Future Generations

27. Value of forage harvested from rangeland by livestock

(Continued)

TABLE 1.3 (Continued)

Criteria and Indicators for Sustainable Rangeland Management

28. Value of production of non-livestock products produced from rangeland
29. Number of visitor days by activity and recreational land class
30. Reported threats to the quality of recreation experiences
31. Value of investments in rangeland, rangeland improvements, and recreation/tourism infrastructure
32. Rate of return on investment for rangeland livestock enterprises
33. Area of rangelands under conservation easements, and number of easements
34. Expenditures (monetary and in-kind) to restoration activities
35. Threat or pressure on the integrity of cultural and spiritual resource values
36. Poverty rate (general)
37. Poverty rate (children)
38. Income inequality
39. Index of social structure quality
40. Community satisfaction
41. Federal transfers by categories (individual, infrastructure, agriculture, etc.)
42. Presence and tenure of natural resource nongovernmental organizations at the local level
43. Sources of income and level of dependence on livestock production for household income
44. Employment diversity
45. Agriculture (ranch/farm) structure
46. Years of education
47. Value produced by agriculture and recreation as of total
48. Employment, unemployment, underemployment, and discouraged workers by the industrial sector
49. Land tenure, land use, and ownership patterns by size classes
50. Population distribution and population change
51. Income differentials from migration
52. Length of residence (native, immigrant > 5 years, immigrant < 5 years)
53. Income by work location vs. residence
54. Public beliefs and attitudes about natural resources

Criterion 5. Legal, Institutional, and Economic Frameworks for Rangeland Conservation and Sustainable Management

55. Extent to which laws and regulations clarify property rights and land tenure arrangements, recognize customary and traditional rights of indigenous people, and provide means of resolving property disputes by due process
56. Nature and extent of institutions and organizations that affect rangeland sustainability
57. Nature and extent of economic policies and practices that affect rangeland sustainability
58. Nature and extent of laws and programs that ensure access to public information and provide opportunities for public participation in land management
59. Nature and extent of natural resource and land management education and assistance programs
60. Nature and extent of land management programs implemented by owners of rangelands

(Continued)

TABLE 1.3 (Continued)

Criteria and Indicators for Sustainable Rangeland Management

61.	Nature and extent of large-scale land planning and assessment projects and related policy review
62.	Nature and extent of rangelands set aside to protect special values
63.	Nature and extent of the various monitoring programs used to evaluate rangeland sustainability
64.	Nature and extent of research and development of new technologies as they affect rangeland sustainability

- providing a starting point for stakeholder dialogue and better informed national policy deliberations; better data, better discussion; and better decisions;
- enhancing interagency cooperation and collaboration and facilitating agency performance planning to shift funding priorities to at-risk areas;
- providing opportunities to better evaluate and improve rangeland management to meet social, economic, and ecological goals; and
- expansion of public awareness and understanding of rangeland sustainability.

Additionally, a broadly accepted set of indicators improves accountability for rangeland stakeholders and Congress through activities, including multiscale, coordinated data reporting; assessing compliance with applicable regulations and laws; and expanding general understanding of rangeland sustainability. Ultimately, effective communication and implementation of criteria and indicators will enhance the quality of debate about rangeland management issues (Wade-Evans and Reuwsaat, 2002).

The first criterion addresses conservation and maintenance of soil and water resources of rangelands. Linking these two resource classes for assessment purposes is practical since soils provide a medium for the capture, storage, and release of water (Whisenanat, 1999). Active flows of water can cause erosion over time and soil erosion has been identified as contributing to human civilization declines over the past 7,000 years (Lowdermilk, 1953). Thus, identification and quantification of rangeland indicators related to soil and water will approximate both the status of rangeland sustainability and associated human communities.

Significantly, the capacity of rangelands to sustainably produce commodities and to satisfy societal values and expectations depends on ecological processes such as soil genesis, nutrient cycling, and energy flow (NRC, 1994). Twelve indicators—six soil-based and five water-based, and one covering both topics—have been identified. Subjects including soil erosion by water and wind, soil organic matter, soil compaction, soil aggregate stability, bare ground, and soil organisms are addressed by soil-based indicators. Biodiversity of aquatic organisms, water quality and chemistry, stream channel geometry, changes in groundwater systems, and duration of flow in intermittent streams are considered in the water-based indicators. The final indicator surfaced by this criterion considers the contamination of rangelands with toxic and/or hazardous materials.

The work of the Sustainable Water Resources Roundtable, which operates as an official Federal Advisory Committee in the United States, ultimately may impact information assessed under this criterion too. Additionally, participants in this criterion group expect the indicators dealing with toxic substances and soil compaction to present special challenges due to the national scale of the overall SRR work relative to the scale at which these indicators are applicable. It appears that these two indicators are extant over small portions of rangelands rather than presenting widespread problems. The question that indicator users must answer is whether an indicator detected over a small geographic area of rangelands can be an effective indicator for national sustainability.

The second criterion quantifies the conservation and maintenance of plant and animal resources of rangelands. The capacity of rangelands to produce commodities and to satisfy societal values and expectations on a sustained basis depends on internal, self-sustaining ecological processes, including the structure and functional dynamics of plant and animal communities (NRC, 1994). As such, SRR identified a need to develop a set of standardized assessment and monitoring methods and protocols to assess the status of plant and animal populations and communities on US rangelands.

It is important to note that some of the generally accepted concepts associated with ecosystem health are related to the maintenance of productive capacity on rangeland ecosystems and the conservation and maintenance of soil and water resources on rangelands. Indicators for productive capacity and soil and water resources have been developed under separate, independent criteria and are not discussed here. Ten indicators have been drafted to capture aspects of ecological sustainability, including broad-based assessment of the location and amount of rangeland in the United States, fire regimes, riparian conditions, fragmentation, invasive species, and species and communities of concern.

Challenges faced by participants in this criterion group included data consistency and standardization issues, as well as gaps in available data sets. Vegetation community definitions, shortcomings of monitoring programs for rare species and specialized habitats, and availability of directly measured genetic diversity information for plants present special problems too.

The third criterion assesses the maintenance of productive capacity on rangeland ecosystems. Rangelands have the capacity to provide the current generation with a wide variety of goods and services as desired by society at a given time. This criterion group currently defines productive capacity as a set of biophysical processes that annually yield plant biomass reflecting the climate, soils, plant composition, and current uses of this biomass. Sustaining productive capacity requires that estimates of this criterion consider temporal and spatial scale issues across a wide variety of goods and services. Maintenance of productive capacity implies that future generations also will be able to obtain their desired mix of market and nonmarket goods from rangelands. It is important to note that productive capacity includes more than forage-based products, such as livestock, and also must address nonconsumptive goods and services, such as wildlife habitat, open space medicinal plants, and wood products. Five indicators have been identified to quantify total acreage, livestock, wildlife, invasive plants, non-forage products, and biomass production.

A difficult question that arises when evaluating rangeland productive capacity is, "Capacity for what?" One rangeland system can produce many goods and services. Some products are mutually exclusive (competitive) while others are compatible (coexisting). However, this is a fundamental principle of all ecosystems; seldom is there a linear exchange between different uses. Multiple species grazing illustrates this concept. Furthermore, identifying and monitoring key goods and services, over long time frames, requires multiscale measurement capability at the ecological site, as well as the ecoregion. Data must be compatible over these long time series. This latter issue has not been clearly defined or coordinated, which provides additional challenges for this criterion group and land managers in general.

The fourth criterion created by the SRR addresses the maintenance and enhancement of multiple economic and social benefits to current and future generations. The ability of rangelands to produce goods and services desired by society is assessed in socioeconomic terms; however, few range-specific measures of social and economic attributes exist. Indeed, social systems have been identified as the least studied component of the rangeland science and management field (Vavra, 1995). The importance of monitoring the social component of sustainability becomes more obvious as difficulties faced by rangeland managers are publicized. Socioeconomic indicators provide a measure of societal values reflecting the allocation of scarce economic resources and illustrate how changes in ecological, legal, and political operations are manifested in economic systems. Twenty-nine indicators have been developed across seven categories, focusing on demographic change in rangeland-dependent areas, community well-being, structure of economic activity, structure of government and social services, character of local social networks, patterns of land tenure, and use-/production-related issues. These issues must encompass both traditional forage-based uses and nontraditional, nonconsumptive uses, such as recreation.

This criterion group has wrestled continuously with questions of scale related to priorities of private and public landowners and aggregation of data. More specifically, issues that are relevant to individual private rangeland owners may not be significant at the county, regional, or national level. Additionally, aggregation of certain data sets may alter outcomes of the analysis. For example, group members expressed concern that consideration of national data from 'rangeland counties' may present inaccurate profiles if information from larger urbanized counties overshadows data derived from less populated, more economically uniform counties. It is possible that this assessment may highlight a need for new research protocols to change the way socioeconomic data are collected and analyzed (Tanaka and Torell, 2002); the SRR criteria and indicators allow experts to definitively resolve this question.

Caveats

The final criterion considered by the SRR deals with legal, institutional, and economic frameworks for rangeland conservation and sustainable management. Indicators within the legal-economic-institutional framework criterion define the extent to which national, state, and local laws, regulations, guidelines, and policy frameworks support conservation and sustainable management of rangelands. Economic indicators complement social measures by assessing changes resulting from adjustments in social, ecological, legal, and political systems. Within this criterion, issues of equity, economic efficacy, cultural traditions, legal rights, and obligations, advancing management theories and skills, greatly influence the long-term sustainability of our nation's rangelands. This criterion contains ten indicators, focusing on the legal, institutional, and economic framework to support conservation and sustainable management, as well as the capacity to measure and monitor changes on rangelands and to conduct and apply research and development to management.

The indicators require the collection and integration of information at county, state, regional, and national scales. Additionally, while data exist to describe institutions, mechanisms, and capacities affecting rangelands, participants in this criterion group have expressed concern about the lack of information addressing how a given institution actually impacts rangeland management. For example, economic policies influencing the use of rangelands can be listed, but the differences a policy may have made in parcel or allotment management cannot be isolated (Lustig, 2002). Since many of these laws, regulations, and policies change over time, tracking the impact of those changes affects how data from the first four criteria can be interpreted.

It is important to note that while Montreal Process and RSF criteria addressing maintenance of ecosystem health and vitality and maintenance of forest contribution to global carbon cycles are not included individually in the SRR list, the indicators that they encompass have been incorporated into the SRR set. Mitchell (2003) asserted that processes including mechanisms of ecosystem recovery, energy and nutrient distribution, and soil degradation, which are subsumed by existing SRR ecological criteria, adequately characterize rangeland health. Thus, the need for a separate rangeland health criterion in the SRR list was negated.

Additionally, rangeland contributions to global carbon cycling are fundamentally different than those of temporal and boreal forests. While forest carbon is stored aboveground, soil organic matter contains the majority of rangeland carbon. Therefore, the SRR soil and water conservation indicator quantifying area and percent of rangeland with significantly diminished soil organic matter and/or high carbon to nitrogen (C:N) ratio offers the rangeland equivalent of these Montreal Process and RSF criteria (Mitchell, 2003).

This assessment marks SRR's first attempt to work with available data sets to populate indicators. Notably, analysts encountered similar challenges to issues itemized in the final draft of the 2003 National Report on Sustainable Forests. The integrated nature of the complementary RSF and SRR criteria and indicator sets left little doubt that priority data issues identified in the RSF draft report could impact the rangeland criteria and indicator effort, too. Particularly relevant priority issues include database management, consistency, and integration; definition of forest and rangeland; implementation of the National Vegetation Classification (NVC) and conversion of existing systems; and human community and economic indicators (USDA Forest Service, 2002).

The existence of data gaps and inconsistent or incomplete information affecting the efficacy of SRR criteria and indicators have been noted already. Generally, the effectiveness of inventory and monitoring systems based on criteria and indicators depends on the resolution of database management, consistency, and integration issues. Authors of the sustainable forest draft report found that little effort has been made to ensure standardization of content, format, and structure of national databases, and critical information about the databases (metadata) is often lacking (USDA Forest Service, 2002). The broad array of databases accessed to populate the SRR indicators selected for this assessment exacerbated this issue, and the data needs and recommendations identified through this assessment are discussed in detail within a dedicated chapter.

The primary concern cited in the final draft of the sustainable forest report focuses on differing operational definitions of forests and rangelands (USDA Forest Service, 2002). SRR participants working in the criterion group addressing conservation and maintenance of plant and animal resources of rangelands also have identified this issue. Currently, overlaps between these definitions results in the addition of total rangeland area plus total forest area summing to more than the existing total land area; the Oregon

Demonstration Project (Goebel et al., 1998) and the 2010 RPA Rangeland Assessment (Reeves and Mitchell, 2011; Reeves and Mitchell, 2012) both found 10% to 15% difference in areas of forests and rangelands depending on definitions. Fortunately, stakeholders from the RSF, the SRR, the Federal Geographic Data Committee (FGDC) Sustainable Forest Data Working Group, FGDC Vegetation Subcommittee, FGDC Sample Inventory and Monitoring of Natural Resources and the Environment Working Group, Bureau of Land Management, Forest Service, Natural Resources Conservation Service, US Geological Survey, Society for Range Management, Society of American Foresters, and NatureServe attempted to better standardize definitions or crosswalk more effectively between data sets. The Oregon Multi-agency Pilot Project, conducted with USDA Forest Service, USDA Natural Resources Conservation Service, and USDI Bureau of Land Management, more fully assessed challenges and solutions. Partners explored how consistent information could be collected to produce unbiased estimates across the landscape using data from the NRI and FIA sampling strategies. Exploratory data analysis was conducted to illustrate some of the uses for the data (Patterson et al., 2014). Consensus on the definitions has not been reached.

Also highlighted is the importance of implementation of the NVC and conversion of existing systems (USDA Forest Service, 2002). NVC is the US federal standard for vegetation classification; however, land management agencies have not yet fully employed it. Efforts related to the NVC to develop standardized methods for designating vegetation community types include, but are not limited to, the Gap Analysis Program, the US Geological Survey/NPS Vegetation Mapping Program, NatureServe, the aforementioned FGDC Vegetation Classification Standards, the ESA Vegetation Classification Panel, major land resource areas (MLRA), Küchler Ecoregions (KE), Environmental Protection Agency ecoregions, the ecological site descriptions being developed by the Natural Resource Conservation Service, and others. Protocols for designating vegetation community types are being developed under each of these programs, and the use of consistent classification systems is essential for the maximization of data-sharing potential (USDA Forest Service, 2002).

Lastly, the sustainable forest draft report called for the development of a national strategy for monitoring human community and economic indicators that includes consideration of social, economic, and cultural impacts of management activities on regions and communities. The culture and economy of resource-dependent communities are linked closely to the lands around them, but information about their cultural and spiritual significance is available only in a qualitative format. While social scientists have techniques to assess values of recreational and aesthetic benefits, methods are lacking to assign quantitative weights to cultural and spiritual values (USDA Forest Service, 2002). Though originally written regarding forests, these points apply for qualitative values of rangeland resources too and add to other challenges already identified by SRR participants working in the criterion group dealing with maintenance and enhancement of multiple economic and social benefits to current and future generations.

Chapters

Subsequent chapters highlight ecological, economic, and social indicators. These chapters also detail adjustments made from the original 64 indicators listed in Table 1.3, to better capture available data used to populate the indicators used in this assessment. A total of 24 social, ecological, and economic indicators are included; this breaks down as 7 sociological indicators, 8 ecological indicators, and 9 economic indicators. However, several of these indicators included multiple components and were subdivided due to data availability and classifications, so a total of 34 metrics is described and depicted in assessment maps, tables, and graphs.

Scale was a consideration throughout the criteria and indicator development process, as well as during the assessment analyses. As a result, indicator data are presented across rangeland systems at multiple spatial scales: counties, ecoregions, MLRA, and congressional districts (CD). Different patterns are visible at the various spatial scales, accentuating the importance of selecting the proper spatial scale for specific questions of interest. Mitchell (2010) noted the need for additional research to further describe how indicators change across scales and how similar indicators at different scales are related.

Various approaches to depicting rangeland sustainability, and the inclusion of ecological, social, and economic elements, exist. Discussion of methods including the intersecting Venn diagram or plates balanced on a fulcrum is reviewed herein. A new approach modeled after a soil texture triangle is also introduced, using a ternary plot algorithm to capture relative influences of social, ecological, and economic components. Results are displayed for 690 rangeland, nonurban counties based on 2017 data.

The Rangeland Sustainability Triangle addresses the challenge of effectively integrating social, economic, and ecological data, which has been a topic of ongoing discussion throughout the SRR's 20-year history. Although the triangle only depicts balance among these three elements within a geographic division, it gives managers and politicians a basis for discussion about the need to equalize influences of these aspects on the county's sustainability. The intent is not to drive value judgments about what combination is good or bad but rather to make available information for decision-making. Additionally, since data available were from a single year, trend analyses were not possible. Better availability of consistent data through time would enhance the assessment of trends in rangeland sustainability at all scales.

Case studies for three regions with interesting assessment outcomes are used to illustrate opportunities identified by indicators. Counties were selected based on their location and MLRA. Regions selected are the Wasatch Front, Central Great Plains, and Northern Great Plains. In each area, indicators point out ecological, social, and economic elements to further explore. For example, in the Wasatch Front, people over 65 years of age and federal transfers may be of interest to county commissioners, as these scores seem low. Social elements stand out as a category for improvement in the Central Great Plains study, to better balance across social, ecological, and economic elements. The Northern Great Plains has opportunities to improve balance by doing more in the ecological and social sectors. Invasive species could be reduced, which in turn could enhance phytomass and production. Ultimately, county commissioners and other decision-makers can look at these indicators and balance across social, ecological, and economic elements to prioritize policy and management actions.

Lessons learned inform the final chapter, which identifies future research needs, data consistency challenges, and key elements still missing that would further improve social, ecological, and economic rangeland sustainability assessment. Shortcomings of included data are discussed. Additionally, data needs are evaluated for the indicators included in this assessment, but also for aspirational indicators that should be included in future assessments to expand the array of reported indicator measures. Given that we reported on 24 indicators out of the original set of 64, over 60% of indicators remain to be populated with data for reporting and discussion. Much work remains to be done.

Methods Used

We developed a process to define rangeland counties and gather data for the indicators. The specific indicators are discussed in detail in Chapters 2 and 3. The processes and methods we used are discussed here to put the succeeding chapters in context. This section just gives a general overview, and each of these will be discussed in more detail throughout the book.

Rangeland and Rangeland Counties

Rangelands were identified using the vegetation types identified by the USNVC (USNVC, 2021). The NVC database was laid over counties. Only those counties that showed greater than 25% rangeland vegetation were used for further analysis. Since there are urban counties that contain more than 25% rangeland, we conducted further analysis. We felt that including these predominantly urban counties would have skewed the economic and social data. To address this issue, we arbitrarily cut population density off at 200 people per acre in a county. Following that analysis, there were a few counties where data were not available or were deemed unreliable. The result was 690 counties in the western United States and Florida deemed rangeland counties.

Data for selected indicators were collected on a county basis. Selected indicators are shown in Chapters 2 and 3. We then looked at the data at different geographical scales. We thought people or organizations may be interested in data for MLRAs, ecoregions, and CD. We aggregated data by estimating the area of each included county in each geographical unit and summed those for that unit. Results were then mapped. Note that not all of the results for each MLRA, ecoregion, or congressional district are shown. For example, if an MLRA only has 10% of its area in a rangeland county, that county's data is used to represent the entire MLRA.

Indicators

For each of the selected indicators, county data were collected from an available source. In order to combine the information in a meaningful way, the data were standardized from 0 to 100, with 100 deemed to be better than 0. We recognize that "better" is a subjective term and may depend on an individual value set. For some indicators, this was straightforward, and others required our interpretation.

Lastly, we calculated means and standard deviations for each standardized indicator. We then calculated ranges of standardized data based on plus or minus 0.5 and 1.5 standard deviations from the mean. Data for each county were then placed into five ranges of standard deviation categories (< –1.5, –1.5 to –0.5, –0.5 to 0.5, 0.5 to 1.5, and >1.5). We refer to those as well below average, below average, average, above average, and well above average. For any county, both the standardized score and standard deviation categories are relative to the other 689 counties. They do not represent how well or bad an individual county is doing in absolute terms.

Assessing Sustainability

The final step was to evaluate rangeland sustainability from a social, economic, or ecological perspective and then all three perspectives together. We have approached this from two different directions. The first is using the standard deviation categories and the second was using a newly developed Sustainability Triangle. This triangle displays the standardized average scores for social, economic, and ecological indicators and the balance among them. The triangle is divided up according to the standard deviation category breaks. Each county (or other geography) set of scores was calculated and placed within the triangle. The areas of the triangle were given names of sustainably balanced, less sustainably balanced, social-economic, social-ecological, ecological-economic, social-oriented, economic-oriented, and ecological-oriented.

Note

1 https://sustainablerangelands.org

Literature Cited

Bartlett, E.T., Maczko, K.A., Mitchell, J.E., 2003. Ecological, economic, and social indicators: assessing rangeland sustainability, in: Allsopp, N. Palmer, A.R., Milton, S.J., Kirkman, K.P., Kerley, G.I.H., Hurt, C.R. (Eds.), *Proceedings of the 7th International Rangeland Congress*, Durban, South Africa, 26 July–1 August 2003. Document Transformation Technolologies, Irene, South Africa, pp. 1897–1908.

Billharz, S., Moldan, B., 1997. Elements of a research agenda, in: Moldan, B., Billharz, S. (Eds.), *Sustainability indicators*.Wiley, Chichester, UK, pp. 389–395.

Coulombe, M.J., 1995. Sustaining the world's forests: the Santiago Agreement. *Journal of Forestry* 93, 18–21.

De Soyza, A.G., Whitford, W.G., Turner, S.J., VanZee, J.W., Johnson, A.R., 2000. Assessing and monitoring the health of western rangeland watersheds. *Environmental Monitoring and Assessment* 64, 153–166.

Dombeck, M., 1998. *A gradual unfolding of a national purpose: a natural resource agenda for the 21st century. Speech to USDA Forest Service employees on 2 March 1998*. US Department of Agriculture Forest Service, Washington, DC.

Fancy, S.G., Gross, J.E., Carter, S.L., 2009. Monitoring the condition of natural resources in US national parks. *Environmental Monitoring and Assessment* 151, 161–174.

Goebel, J.J., Shreder, H.T., House, C.C., Geissler, P.H., Olsen, A.R., Williams, W.R., 1998. Integrating surveys of terrestrial natural resources: the Oregon demonstration project. Inventory and Monitoring Report No. 2, US Department of Agriculture Forest Service, Washington, DC.

Graetz, D., 1994. Grasslands, in: Meyer, W.B., Turner II, B.L. (Eds.), *Changes in land use and land cover: a global perspective*. Cambridge University Press, Cambridge, pp. 125–147.

[Heinz] The H. John Heinz III Center for Science, Economics, and the Environment, 2002. *The state of our nation's ecosystems: measuring the lands, waters, and living resources of the United States*. Cambridge University Press, New York, NY.

Heintz, T., 2002. The roles and importance of sustainability indicators, in: Bartlett, E.T., Mitchell, J.E. (Eds.), Sustainable rangeland management: status of a roundtable to determine criteria and indicators, a work in progress. Unpublished proceedings of a symposium sponsored by the *Sustainable Rangelands Roundtable at the 2002 Annual Meeting of the Society for Range Management*, Kansas City, pp. 7–14. http://www.sustainablerangelands.org/wp-content/uploads/2021/05/2002SRMProceedings_Searchable.pdf

Linestone, H.A., Turoff, M. (Eds.), 1975. *The Delphi method: techniques and applications*. Addison-Wesley, Reading, MA.

Lowdermilk, W.C., 1953. *Conquest of the land through 7,000 years*. US Department of Agriculture, Natural Resources Conservation Service Agricultural Information Bulletin Number 99, Washington, DC.

Lustig, T.D., 2002. Status of institutional criteria and indicators, in: Bartlett, E.T., Mitchell, J.E. (Eds.), Sustainable rangeland management: status of a roundtable to determine criteria and indicators, a work in progress. Unpublished proceedings of a symposium sponsored by the *Sustainable Rangelands Roundtable at the 2002 Annual Meeting of the Society for Range Management*, Kansas City, pp. 59–62.

[MEA] Millennium Ecosystem Assessment, 2005. *Ecosystems and human well-being*. Island Press, Washington, DC.

Miller, A., Cuff, W., 1986. The Delphi approach to the mediation of environmental disputes. *Environmental Management* 10, 321–330.

Mitchell, J.E., 2003. The sustainable rangelands roundtable: a journey toward sustainable management, in: Allsopp, N. Palmer, A.R., Milton, S.J., Kirkman, K.P., Kerley, G.I.H., Hurt, C.R. (Eds.), *Proceedings of the 7th International Rangeland Congress*, Durban, South Africa, 26 July–1 August 2003. Document Transformation Technolologies, Irene, South Africa.

Mitchell, J.E. (Ed.), 2010. *Criteria and indicators of sustainable rangeland management*. University of Wyoming Extension Publication Number SM-56. Laramie, WY. 227 p.

[NRC] National Research Council, 1994. *Rangeland health: new methods to classify, inventory, and monitor rangelands*. The National Academies Press, Washington, DC. https://doi.org/10.17226/2212.

[NRC] National Research Council, 2000. *Ecological indicators for the nation*. The National Academies Press, Washington, DC. https://doi.org/10.17226/9720.

[NRCBSD] National Research Council Board on Sustainable Development, 1999. *Our common journey: a transition toward sustainability*. National Academy Press, Washington, DC.

Patterson, Paul L., Alegria, J., Jolley, J.L., Powell, D., Goebel, J.J., Riegel, G.M., Riitters, K.H., Ducey, C., 2014. Multi-agency Oregon Pilot: working towards a national inventory and assessment of rangelands using onsite data. General Technical Report RMRS-GTR-317. US Department of Agriculture, Forest Service, Rocky Mountain Research Station, Fort Collins, CO.

Pellant, M., Shaver, P., Pyke, D.A., Herrick, J.E., 2005. *Interpreting indicators of rangeland health*. Tech. Ref. 1734-6, Version 4. US Department of the Interior, Bureau of Land Management National Science and Technology Center, Denver, CO. BLM/WO/ST-00/001+1734/REV05, p. 122.

Reeves, M.C., Mitchell, J.E., 2011. Extent of coterminous US rangelands: quantifying implications of differing agency perspectives. *Rangeland Ecology and Management* 64, 585–597. https://doi:10.2111/REM-D-11-00035.1.

Reeves, M.C., Mitchell, J.E., 2012. *A synoptic review of U.S. rangelands: a technical document supporting the 2010 USDA Forest Service RPA Assessment*. General Technical Report RMRS-GTR-288, US Department of Agriculture, Forest Service, Rocky Mountain Research Station, Fort Collins, CO.

Rowe, H.I., 2002. A new role for delphi? Refining roundtable progress. *Abstracts, 55th Annual Meeting of the Society for Range Management*, Kansas City, MO, USA, 13–19 February 2002, Society for Range Management, Denver, CO, p. 71.

Shields, D., Bartlett, E.T., 2002. Applicability of Montreal Process Criterion 6 - long-term socio-economic benefits to rangeland sustainability. *International Journal of Sustainable Development and World Ecology* 9(2), 95–120.

Tanaka, J.A., Torell, L.A., 2002. Status of economic criteria and indicators for the sustainable rangelands roundtable, in: Bartlett, E.T., Mitchell, J.E. (Eds.), Sustainable rangeland management: status of a roundtable to determine criteria and indicators, a work in progress. Unpublished proceedings of a symposium sponsored by the *Sustainable Rangelands Roundtable at the 2002 Annual Meeting of the Society for Range Management*, Kansas City, pp. 51–53.

[SRR] Sustainable Rangelands Roundtable, 2002. Criteria and indicators for sustainable rangelands – executive summary. https://www.sustainablerangelands.org/reports/2003Report/SRRExecSumm.pdf (accessed 30 January 2021).

[TNC] The Nature Conservancy, 1996. *Conservation by design: a framework for mission success*. The Nature Conservancy, Arlington, VA.

[USNVC] United States National Vegetation Classification, 2021. The U.S. national vegetation classification, your guide to inventorying natural and cultural plant communities. http://usnvc.org/ (accessed 25 January 2021).

[UNCED] United Nations Conference on Environment and Development, 1992a. *Statement of forest principles*, United Nations, Rio de Janeiro, Brazil, 3–14 June 1992. https://www.un.org/en/conferences/environment/rio1992.

[UNCED] United Nations Conference on Environment and Development, 1992b. *Agenda 21*, United Nations, Rio de Janeiro, Brazil, 3–14 June 1992. https://www.un.org/en/conferences/environment/rio1992.

USDA Forest Service, 2002. *National report on sustainable forests – 2003*, Final Draft, US Department of Agriculture, Forest Service, Washington, DC.

Vavra, M., 1995. Rangeland research: strategies for providing sustainability and stewardship to the rangelands of America and the world. *Rangelands* 17(6), 200–214.

Wade-Evans, J., Reuwsaat, T., 2002. Value of sustainability indicators for rangeland management and policy, in: Bartlett, E.T., Mitchell, J.E. (Eds.), Sustainable rangeland management: status of a roundtable to determine criteria and indicators, a work in progress. Unpublished proceedings of a symposium sponsored by the *Sustainable Rangelands Roundtable at the 2002 Annual Meeting of the Society for Range Management*, Kansas City, pp. 15–16.

Whisenanat, S.G., 1999. *Repairing damaged wildlands: a process-oriented, landscape-scale approach*. Cambridge University Press, Cambridge.

[WCED] World Commission on Environment and Development, 1987. *Our common future*. Oxford University Press, New York.

[WSSD] World Summit on Sustainable Development. 2002. *Plan of implementation*. World Summit on Sustainable Development, Johannesburg. https://www.un.org/en/conferences/environment/johannesburg.

2

Ecological Indicators

Numerous ecological indicators are available today that were not available even ten years ago. Indicators such as rangeland vegetation productivity and phytomass are now available and will provide needed context for rangeland assessment. In addition to performance metrics, basic measures of rangeland area are also now available for the coterminous United States at a scale of 30 m spatial resolution. Using these area metrics, the amount of fragmentation of rangeland landscapes can also be determined. In addition, fragmentation metrics and evaluations of vegetation structure and composition also influence the present vegetation departure (VDep) from hypothesized historic ranges. For example, consider the Snake River floodplain where cheatgrass is pervasive. Many of these sites have burned more than five times since 1984. This information is critical to include in this rangeland assessment and is now available. Collectively, these and other indicators offer unprecedented ability to use indicators of the sustainability of rangelands from an ecological perspective in a way that has not heretofore been possible.

RANK SCORING

For all social, economic, and ecological indicators, raw data for each indicator were ranked among the 690 rangeland counties. Counties are categorized by how many standard deviations they are from the mean of all counties. All analysis and discussion in this chapter are based on counties, ecoregions, MLRAs, or congressional districts (CD) compared to all of the other counties, ecoregions, MLRAs, or CD, respectively, using the standard deviation categories.

Ecological indicators are derived from three SRR criteria as indicated in Chapter 1. We do not divide them by the criteria for purposes of descriptions. Table 2.1 shows the original indicator developed by the Roundtable, how the description was changed due to the availability of data, and the data source. While the revised indicator has changes in it, the intent of the original indicator is retained as much as possible. Each indicator will be discussed in the following sections.

DOI: 10.1201/9781003279747-2

TABLE 2.1

Ecological Indicators Used in the National Assessment

Original Indicator	Revised Indicator	Data Source
11. Extent of land area in rangeland	Extent of land area in rangeland	Reeves and Mitchell (2011)
14. Fragmentation of rangeland and rangeland plant communities	Fragmentation of rangeland and rangeland plant communities	Reeves et al. (2018)
12. Rangeland area by vegetation community	Rangeland area by vegetation community	Rollins (2009); Reeves (2016)
18. Area of infestation and presence/absence of invasive and non-native plant species of concern	Area of infestation and presence/absence of invasive and non-native plant species of concern	Center for Invasive Species and Ecosystem Health (2020)
18. Area of infestation and presence/absence of invasive and non-native plant species of concern	Area of infestation and presence/absence of major invasive and non-native plant species of concern	Center for Invasive Species and Ecosystem Health (2020)
21. Rangeland aboveground biomass	Rangeland aboveground phytomass	Rollins (2009); Reeves (2016)
22. Rangeland annual productivity	Rangeland annual productivity	Reeves and Lankston (2020)
23. Number of domestic livestock on rangeland	Number of domestic livestock on rangeland	NASS (2019)
7. Percent of surface water on rangeland areas with significant deterioration of their chemical, physical, and biological properties from acceptable levels	Percent of surface water on rangeland areas with significant deterioration of their chemical, physical, and biological properties from acceptable levels	EPA (2020)

Indicator Descriptions

Extent of Land Area in Rangeland

The *extent of land area in rangeland* stems from the work of Reeves and Mitchell (2011), who developed spatially explicit data describing the extent of rangelands according to differing agency perspectives. The rangeland extent reported here for each spatial unit corresponds to the rangeland definition[1] adopted by the USDA Natural Resources Conservation Service (NRCS, 2021) National Resources Inventory (NRI). This definition gives rise to larger rangeland area estimates since the NRI has a more liberal view of what constitutes "rangeland" compared with the USDA Forest Service, which effectively classifies any land supporting trees in excess of 10% canopy cover or 5% canopy cover in the case of woodland species. On one hand, these issues seem merely academic but nevertheless result in dramatically different characterizations of land cover types in the coterminous United States. In total, the NRI view leads to an estimate of rangelands of about 662 million acres, while the USDA Forest Service view yields about 601 million ac with most of the differences occurring in woodlands and in lands where historical fire regimes restricted trees to low densities (e.g., Florida flatwoods). The nuances of these definitions are evident in Reeves and Mitchell (2011). The work by Reeves and Mitchell (2011) identifies and maps three classes of rangelands relevant here, including (1) rangeland, (2) transitional rangeland, and (3) afforested rangeland. For this sustainability assessment, the "rangeland" and "afforested rangeland" categories are used to identify the extent of rangeland. Transitional rangelands are those that meet most of the criteria of rangelands (e.g., dominated by herbaceous or shrub species) but will succeed to a forest with time. These transitional rangelands are most commonly found where forest thinning or clear-cutting has removed tree overstory and reduced tree density, enabling herbaceous and shrub species to become dominant but only for a short period of time. Conversely, afforested rangelands are those lands that historically exhibited less than 25% canopy cover by trees but presently, possibly owed to fire exclusion, exhibit more than 25% cover by trees. Examples of this would be areas in the western United States experiencing conifer encroachment. Likewise, the southeastern United States historically (pre-Euro-American settlement) experienced summertime fire regimes producing very low tree cover and savanna-like scenarios thereby meeting the criteria for "rangeland." However, due to dramatic increases in tree density, these are characterized as "afforested rangelands" and included in the rangeland area evaluation for this assessment.

Rangeland Area by Vegetation Community

The *rangeland area by vegetation community* was evaluated using the VDep product from the Landfire product suite (Rollins, 2009). In the Landfire system, VDep indicates how different current vegetation on a landscape is from estimated historical conditions. VDep is based on changes to species composition, structural stage, and canopy closure, relative to the historical conditions. For a given biophysical setting (Rollins, 2009; in this case rangeland sites), the range of pre-Euro-American settlement conditions is presumed to be known and defined in the Landfire Project. Likewise, the present-day vegetation conditions are presumed to be known through other spatially explicit data including existing vegetation type (EVT), existing vegetation cover (EVC), and existing vegetation height (EVH). By comparing historical conditions (pre-Euro-American settlement) to present-day vegetation structure and composition, departure can be estimated. For example, consider a tallgrass prairie biophysical setting in which historically woody species were constrained to less than 10% cover. If that site presently exhibits woody cover by a shrub or tree species, such as eastern red cedar (*Juniperus virginiana*) in excess of 10%, the site would be considered departed. This method is sensitive to fire regime changes but also to invasive species, which indeed may be influenced by fire feedback cycles, as in the case of cheatgrass (Balch et al., 2012).

Area of Infestation and Presence/Absence of Invasive and Non-native Plant Species of Concern

Invasive species presence was quantified using the citizen science observations from the Bugwood.org website managed by the Center for Invasive Species and Ecosystem Health (2020). It is not possible to consistently estimate the area of infestation across all lands. Although quantifying the presence of invasive species is more tractable, it is not easy. The presence of invasive and non-native plant species of concern was quantified in this assessment in two ways. First, the total number of invasive species tallied for a given unit of investigation (i.e., counties or counties combined with Küchler Subsections) was estimated. Second, the presence of key exotic species that are dramatically changing the ecological status of landscapes and influencing fire regimes, resistance to drought, etc., was estimated. These species are widely recognized as being highly detrimental to many western landscapes, and their presence inarguably leads to relatively lower sustainability rankings. The highly influential species we evaluated for presence or absence included cheatgrass (*Bromus tectorum*), red brome (*Bromus rubens*), knapweeds and starthistles (*Centaurea* spp.), ventenata (*Ventenata dubia*), leafy spurge (*Euphorbia esula*), and salt cedar (*Tamarisk* spp.). The metric used in the case of the influential species is a count of the number of these species that occur in a specific geography. It follows that the total number or the worst-case scenario would be finding all six of these species in a given summary unit.

Fragmentation of Rangeland and Rangeland Plant Communities

Fragmentation of rangeland and rangeland plant communities is quantified using the work by Reeves et al. (2018), who evaluated stressors on rangelands westwide, including fragmentation, exurban development, and human modification. Methods and theories for quantifying fragmentation abound, but here fragmentation is defined as the Euclidean distance between human-modified landscape patches. Human modification as described in Crooks et al. (2017) and evaluated in Reeves et al. (2018) was represented as spatially explicit data describing the relative amount of human-caused disturbance in 90 m by 90 m pixel. Human modification was estimated using a regression tree approach. Air photos were evaluated at 6,000 locations, and visible human modifications were digitized. These digitized polygons were used to estimate the percent of modification present within each scene. The amount of modification was extrapolated using a regression tree with numerous abiotic and biotic predictors outlined in Reeves et al. (2018). Using this approach, modification was estimated on a 90 m grid spatial resolution in a continuous ramp from 0 (no disturbance) to 100 (full disturbance). In this assessment, a threshold of 50% modification was used to quantify fragmentation. In other words, to be considered in the Euclidean distance formulation, a cell must be greater or equal to 50% modified. In this manner, a greater distance between modified cells (raster cells) suggests greater intactness or less fragmentation. Obviously, if a lower threshold were chosen (e.g., 30% modification), greater estimates of fragmentation would result, so the metric is sensitive to the human modification threshold.

Rangeland Aboveground Phytomass

Estimates of *rangeland aboveground phytomass* are now available for use in national assessments like this. The Rangeland Vegetation Simulator (Reeves, 2016) contains a large suite of allometric equations linking cover, height, and species information to standing carbon and therefore biomass. Using spatially explicit estimates of vegetation cover, height, and composition (e.g., EVC, EVH, and EVT from Landfire) standing biomass can be estimated in shrubs for all coterminous US rangelands.

However, it is unclear how to interpret phytomass. For example, consider a rangeland site that is exhibiting 10,000 pounds per acre of standing phytomass. Is that a positive or a negative attribute for rangeland sustainability? Does less phytomass indicate less sustainable situations? Is the converse true? From a fire behavior and risk perspective, when increasing phytomass by large statured shrubs, such as Utah juniper (*Juniperus utahensis*) or basin big sagebrush (*Artemisia tridentata* spp. *tridentata*), the increase in phytomass is not a positive scenario. Likewise, limits to site productivity suggest that excess phytomass may reduce site diversity and forage (Coultrap et al., 2008), although the effects are not universally consistent (Miller et al., 2000). Perhaps a better way of evaluating the impact of phytomass on rangeland sustainability is to compare the observed phytomass with expected biomass based on desired future conditions. At this time, quantifying desired future conditions for counties of the United States is beyond the scope of this book.

Rangeland Annual Productivity

The *Rangeland Annual Productivity* measure used in this assessment emanates from the Rangeland Production Monitoring System (Reeves and Lankston, 2020) representing annual production on coterminous US rangelands from 1984 to 2020. Using this unprecedented time series of annual production for

US rangelands, we developed two metrics associated with annual productivity. First, we quantified the trend in annual productivity using a simple linear correlation of annual production (pounds per acre) with respect to time (1984–2020). Second, we quantified the average productivity. The average production is an indicator of the amount of forage available, on average, in a given year, suggesting the number of animal units that can be sustainably managed on the landscape.

Although average production is a useful metric for evaluating productivity, others such as trends through time and interannual variability can be used for evaluating sustainability. There is a significant asymmetry that has occurred in the temporal trend of annual production from 1984 to 2020. In the southwestern United States, particularly in the Four Corners region, productivity has been declining since 1984. In contrast, the more northerly rangeland extents, such as in the Pacific Northwest and Northern Great Plains, have been experiencing increases in annual production. It is notable that these trends that have been occurring since 1984 mirror those of climate change studies, suggesting growing conditions will deteriorate in the future in more southern rangelands and improve or maintain in the north. This is a significant finding, suggesting that the decreasing annual production in the south will likely continue in the future, indicating a relatively less sustainable situation when compared with the more northern rangelands. Although we did not include the analysis of trends in this assessment, it does add context for interpreting results.

Number of Domestic Livestock on Rangeland

The *number of domestic livestock on rangeland* (beef cattle, sheep, and meat goats) present is an indication of the productive capacity of rangelands. While other uses of rangelands or drought can reduce the number of livestock present on rangeland, at any point in time, the total numbers can represent what the rangelands can sustain. Livestock numbers come from more generalized National Agricultural Statistics Service (NASS, 2019) survey data, so we do not know what proportion of livestock use rangelands nor for how long. Since the counties we use are rangeland counties as described in Chapter 1, we assume there is a correlation between livestock numbers and their use of rangelands. We can also assume there is a high correlation between numbers of livestock and other indicators such as phytomass and annual productivity.

Percent of Surface Water on Rangeland Areas with Significant Deterioration of Their Chemical, Physical, and Biological Properties from Acceptable Levels

The *percent of surface water on rangeland areas with significant deterioration of their chemical, physical, and biological properties from acceptable levels* uses the impaired waters list as the only water indicator that we are aware of that is consistently and comprehensively available. These data are gathered by individual states to comply with Section 303(d) of the Clean Water Act (EPA, 2020). These spatially explicit data describing impaired waters are presently only available in vector format describing the length of the impaired reach. As a result, each line segment (impaired water reach) was summed for each land unit evaluated. No weighting was given to the width or volume of a given reach. For example, a small stream receives the same weighting as a large river.

Results

It is important to clarify the relationship between rangeland health and the indicators of rangeland sustainability used in this assessment. It is tempting to assume that the ecological indicators are synonymous with rangeland health or rangeland condition. We might agree that healthy rangelands are more likely to sustainably produce goods and services. For example, cattle numbers are not an ecological indicator of rangeland health, but they are an ecological indicator of rangeland sustainability.

Extent of Land Area in Rangeland

As expected, the *extent of land area in rangeland* decreases from west to east, reflecting both the decreasing size of counties and an increasingly fragmented landscape toward the eastern edge of the rangeland extent (Figure 2.1). The southwestern part of the coterminous United States exhibits the greatest rangeland areas. This pattern also follows a general trend toward increasingly larger counties from east to west, so it is useful to also consider the proportion of land in a county that is considered rangeland. Figure 2.1 indicates that there are three regions where the percentage of land that is by rangeland is relatively high, including the Northern Great Plains, Northern Great Basin, and the southwestern United States. The extent of rangeland area somewhat follows patterns of fragmentation as well.

Fragmentation of Rangeland and Rangeland Plant Communities

As expected, the *fragmentation of rangelands and rangeland plant communities* decreases from west to east, reflecting increasing private landownership and increasing diversity of other land use classes, such as pastureland and crops (Figure 2.2). When interpreting Figure 2.2, it is important to study how fragmentation is quantified and portrayed. The fragmentation data from Theobald (2013) and further processed in Reeves et al. (2018) are offered as

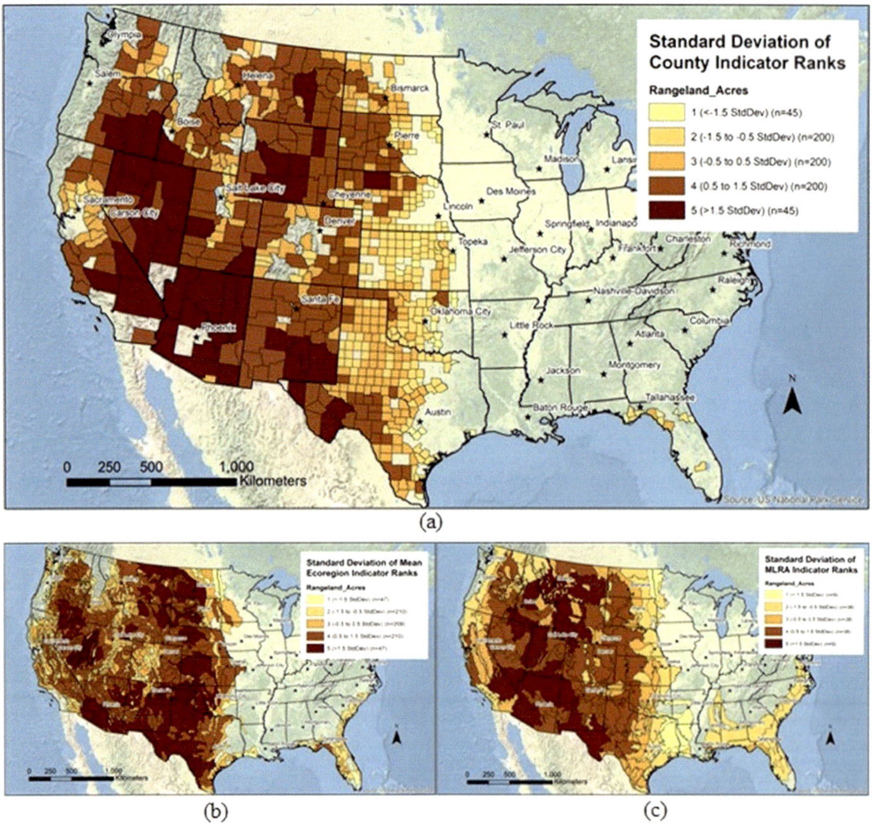

FIGURE 2.1
Rangeland acres by (a) County, (b) Ecoregion, and (c) MLRA.

Euclidean distance between human-modified patches. Thus, increasing distances between modified patches indicate less fragmentation (more intact landscapes). Decreasing fragmentation is typically viewed as a positive attribute of the landscape. As a result, the areas with decreased fragmentation (larger distance between modified patches) are shown with a positive sign because this indicates a positive attribute of rangeland landscapes. Similar to rangeland extent, the least fragmented landscapes occur in the southwestern part of the coterminous United States and the Great Basin region.

Rangeland Area by Vegetation Community

Rangeland area by vegetation community can be evaluated through numerous viewpoints. However, comprehensive and consistent vegetation surveys

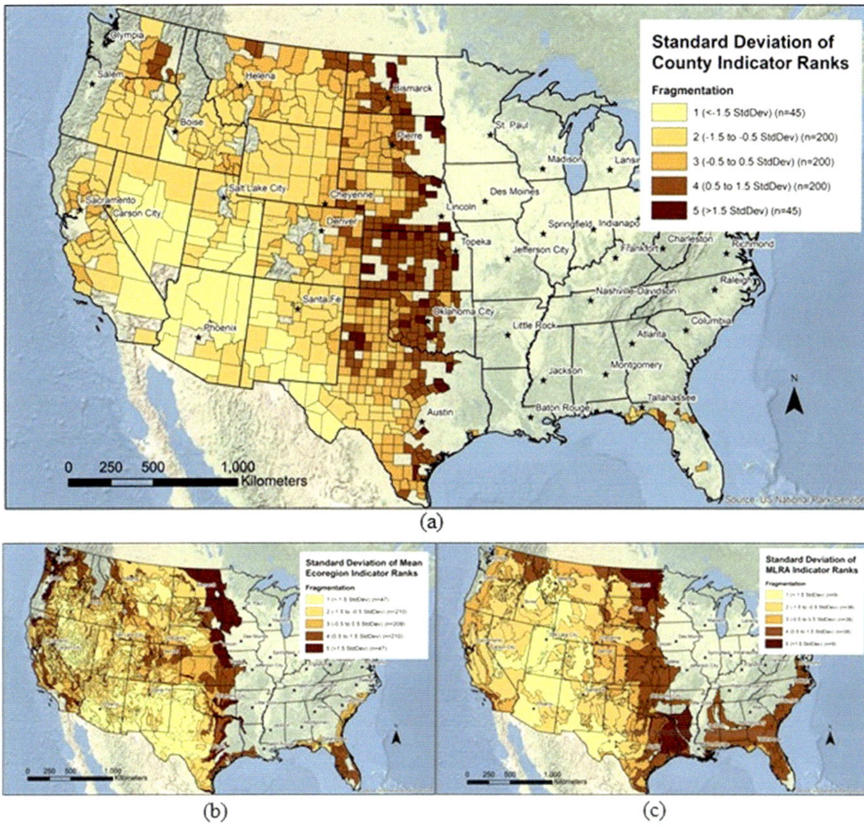

FIGURE 2.2

Fragmentation of rangeland and rangeland plant communities by (a) County, (b) Ecoregion, and (c) MLRA.

are not conducted across all ownerships in the coterminous United States. As a result, we resorted to a geospatial assessment of rangeland area by vegetation community. It is not clear what advantage could be gained by simply tabulating rangeland area by vegetation community without further distinction and context. For example, assume we know that a county in the Great Basin is dominated by equal proportions of vegetation classes including Intermountain Basins Big Sagebrush Shrubland, Intermountain Basins Mixed Salt Desert Scrub, Intermountain Basins Greasewood Flat, Intermountain Basins Big Sagebrush Steppe, and Great Basin Xeric Mixed Sagebrush Shrubland Ecological Systems (Comer et al., 2003). This begs the question, "How can this information be used in a sustainability assessment?"

To use this kind of information, more site-specific evaluations would be necessary to leverage this kind of information to evaluate the relative usefulness of various vegetation types for assessing rangeland sustainability. In an effort to make a more useful and quantitative metric regarding rangeland area by vegetation community, we chose to implement the VDep product from the Landfire Program (Rollins, 2009). This departure metric is more of an ecological indicator than it is a production indicator. For example, a rangeland vegetation community can be highly departed but still provide forage for domestic and wild ungulates, as well as potentially maintaining some aspects of rangeland health. The VDep metric identifies areas where the composition and structure of the current vegetation are dissimilar to pre-Euro-American settlement states. In this manner, the VDep is quite sensitive to non-native invasive species. As a result, areas of high departure include the Snake River floodplain and Northern Great Basin, the California annual grasslands, and eastern Washington and parts of eastern Oregon (Figure 2.3). Likewise, departure can occur when succession classes (state class of vegetation, including vegetation structure and composition) are uncharacteristic. Easily understandable situations meeting these criteria would be the significant brush encroachment that has occurred across much of southeastern Arizona and the eastern edge of the tallgrass prairie, where trees are experiencing a significant increase relative to pre-Euro-American settlement conditions owing primarily to fire return intervals that are longer than previously experienced. Counties with the highest levels of departure include Richland and Ransom Counties in eastern South Dakota, and Culberson, Loving, and Ward Counties in Texas. Large areas of western Texas, eastern Washington, and central California also exhibit significant departure (Figure 2.3). In contrast, the Desert Southwest and Northern Great Plains exhibit relatively less departure.

Area of Infestation and Presence/Absence of Invasive and Non-native Plant Species of Concern

As with fragmentation, higher standard deviations do not indicate higher counts of invasive species, and the presence of fewer invasive species is seen as a positive aspect of rangeland sustainability. Figure 2.4 indicates that the counties with the highest counts of invasive species generally occur in the Pacific Coast region. Invasive species counts increased from east to west across the extent of coterminous US rangelands. Overall, California has the greatest number of counties with very high counts of invasive species. Butte County in California has the highest recorded number of invasive species with 518 estimated to occur, while several counties in Texas exhibit fewer than 5. In particular, the Southern Great Plains exhibit the lowest counts of invasive species. The relatively high number of invasive species on the West Coast may be a result of suitable growing seasons given the Mediterranean climate dominating the area. In addition, invasive species observations in the

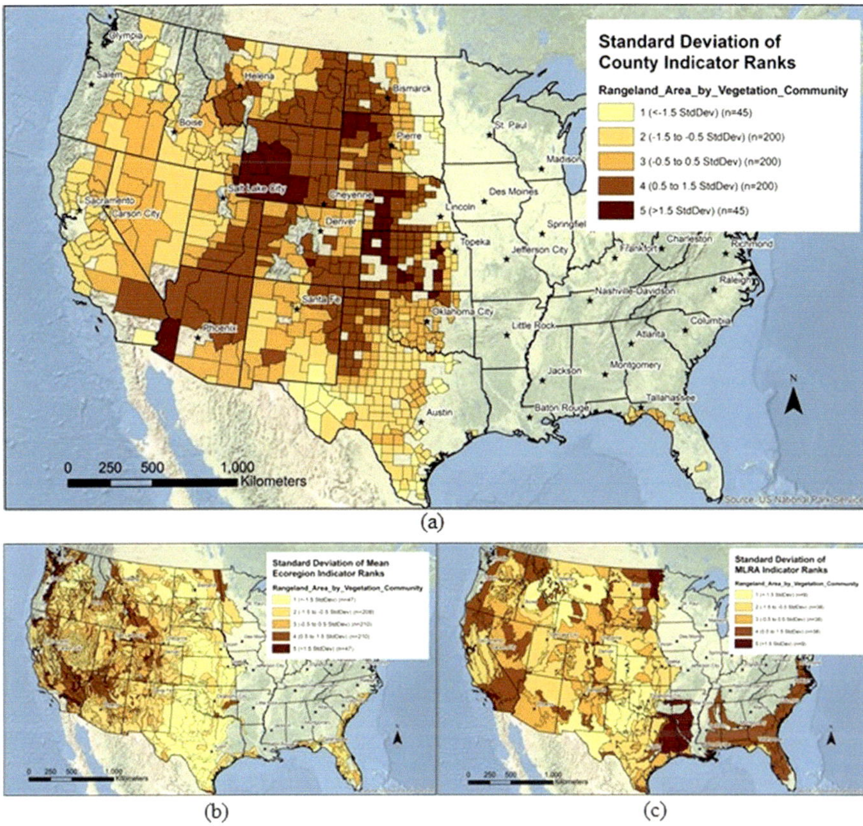

FIGURE 2.3
Rangeland area by vegetation community Represented by the Deviation from Historical Vegetation by Rangeland Area by (a) County, (b) Ecoregion, and (c) MLRA.

Bugwood database are part of a citizen science project typically reported by volunteers. The fact that the Bugwood data set is populated by volunteers may also influence the patterns observed in Figure 2.4. Thus, it is possible that the invasive species counts in the Great Plains may be higher than suggested in Figure 2.4 because participation in the Bugwood program may be lower in these areas than it is on the West Coast.

With some exceptions, the pattern of particularly harmful invasive species corresponds with the total count shown in Figure 2.5. West of the longitudinal line running through Denver, Colorado, almost every county contains at least four of the six particularly harmful invasive species. As with the total count of invasives mentioned earlier, this may reflect climate and growing conditions that could be favorable for this suite of selected key species. The ability of landscapes across the western United States to resist invasion,

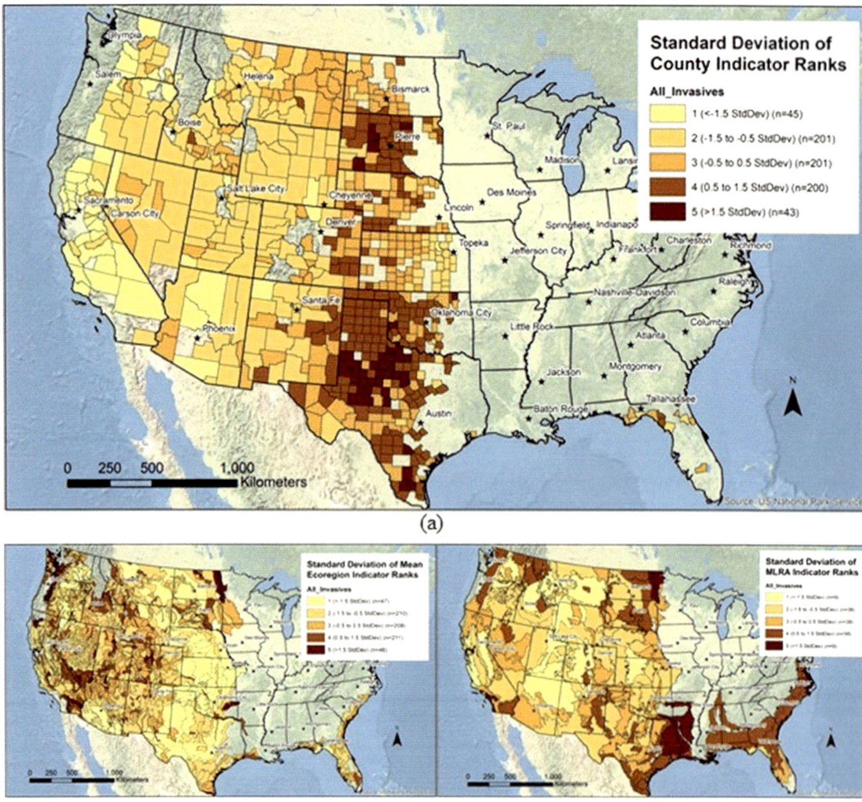

(b) (c)

FIGURE 2.4
Area of infestation and presence/absence of invasive and non-native plant species of concern by (a) County, (b) Ecoregion, and (c) MLRA.

especially by annual grasses such as cheatgrass, is highly variable and often strongly determined by soil temperature and moisture regimes. Warmer and drier soil types tend to have less resistance to invasion, which may also explain a greater instance of these highly influential invasives in more arid regions. Resistance may be influenced by the choice of species to evaluate in this assessment, or it may reflect asymmetric contribution rates to the Bugwood program.

Rangeland Aboveground Phytomass

The southern part of the rangeland domain exhibits the greatest amount of *Rangeland Aboveground Phytomass* (Figure 2.6). Madera County in California and Pima County in Arizona exhibit the greatest amount of

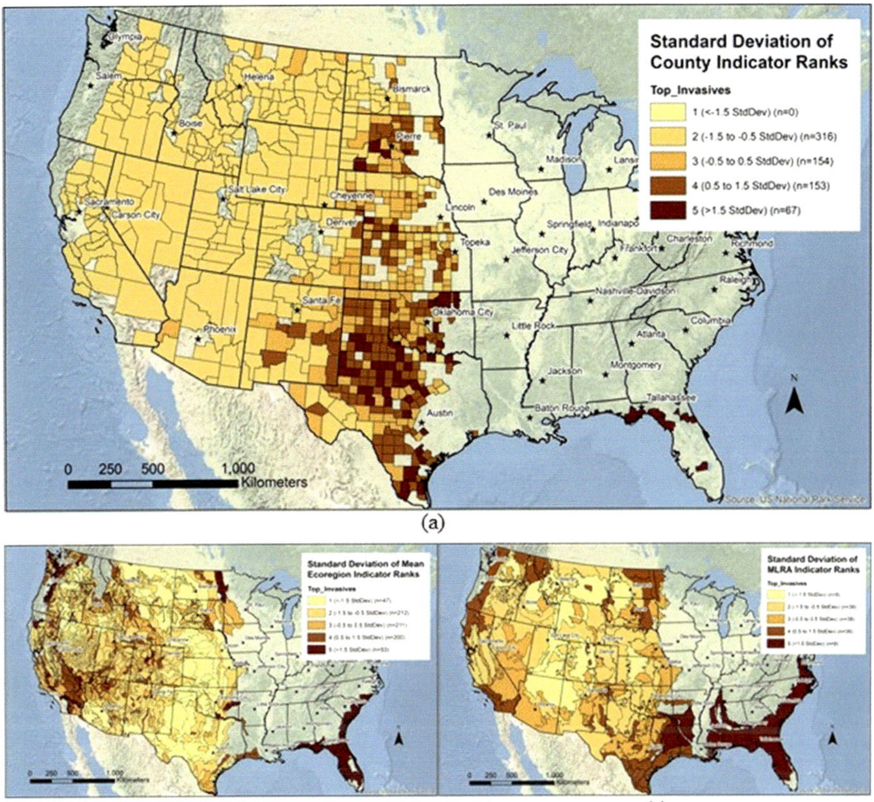

(a)

(b) (c)

FIGURE 2.5
Area of infestation and presence/absence of major invasive and non-native plant species of concern by (a) County, (b) Ecoregion, and (c) MLRA.

standing phytomass, each with approximately 12,000 pounds per acre of total phytomass. West-central Texas, central California, and southeastern Arizona are areas with relatively high phytomass values as well. From south to north, across the extent of US rangelands, a decreasing gradient of phytomass exists.

Rangeland Annual Productivity

Rangeland annual productivity and *rangeland annual phytomass* are linked. However, unlike phytomass, there are clear and intuitive uses for annual production since decreasing annual production over a long period of time indicates an unsustainable situation, especially when normalized against like vegetation types (e.g., comparing similar sites in the shortgrass steppe

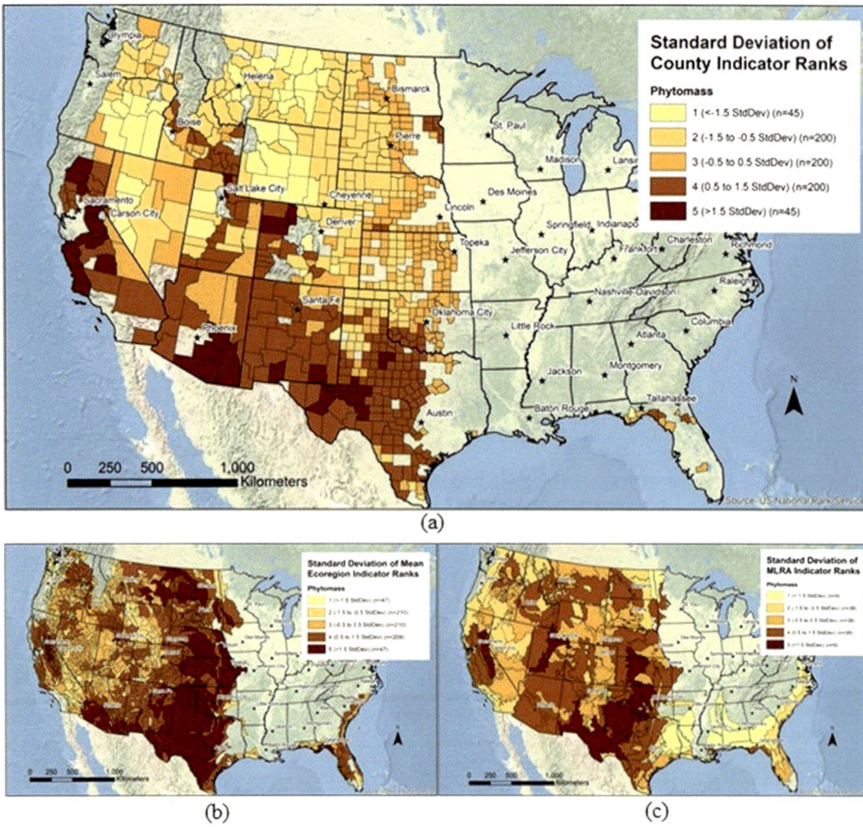

FIGURE 2.6
Rangeland aboveground phytomass by (a) County, (b) Ecoregion, and (c) MLRA.

domain on the Central Great Plains). In this assessment, we used the average annual production across 37 years as the indicator of sustainability. Higher annual production suggests greater potential to achieve a more varied economic system, a more positive attribute of rangeland sustainability. Figure 2.7 indicates a strong gradient of increasing annual production from west to east, reflecting greater effective moisture toward the eastern edge of the coterminous US rangeland extent. Average annual production ranges from 120 to 5,140 pounds per acre in Esmerelda County, Nevada, and Lafayette, Liberty, and Flagler Counties in Florida, respectively.

Number of Domestic Livestock on Rangeland

The *number of domestic livestock on rangelands* (Figure 2.8) somewhat follows the annual productivity map (Figure 2.7). There are significant differences

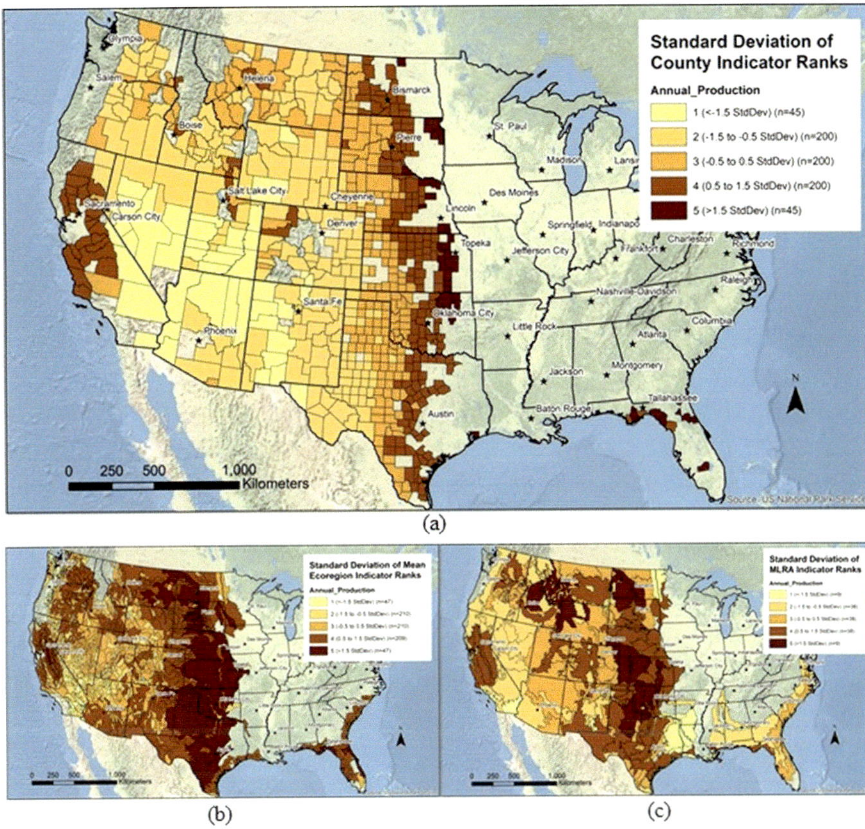

FIGURE 2.7
Rangeland annual productivity by (a) County, (b) Ecoregion, and (c) MLRA.

that are especially apparent in the Northern Great Plains and the Great Basin. This can be explained somewhat because numbers are shown by counties; larger counties in the Great Basin, Montana, and Wyoming have more area to raise livestock, even though they have lower annual vegetation productivity. Relative to the Central and Southern Great Plains, these counties rank higher, even though as a region they rank lower, as shown by the ecoregions (Figure 2.8b), indicating that livestock numbers are relatively higher throughout the Great Plains.

Percent of Surface Water on Rangeland Areas with Significant Deterioration of Their Chemical, Physical, and Biological Properties from Acceptable Levels

Figure 2.9 reveals a nearly random pattern. It is difficult to derive meaningful information from the 303(d) impaired water data as it was used in this

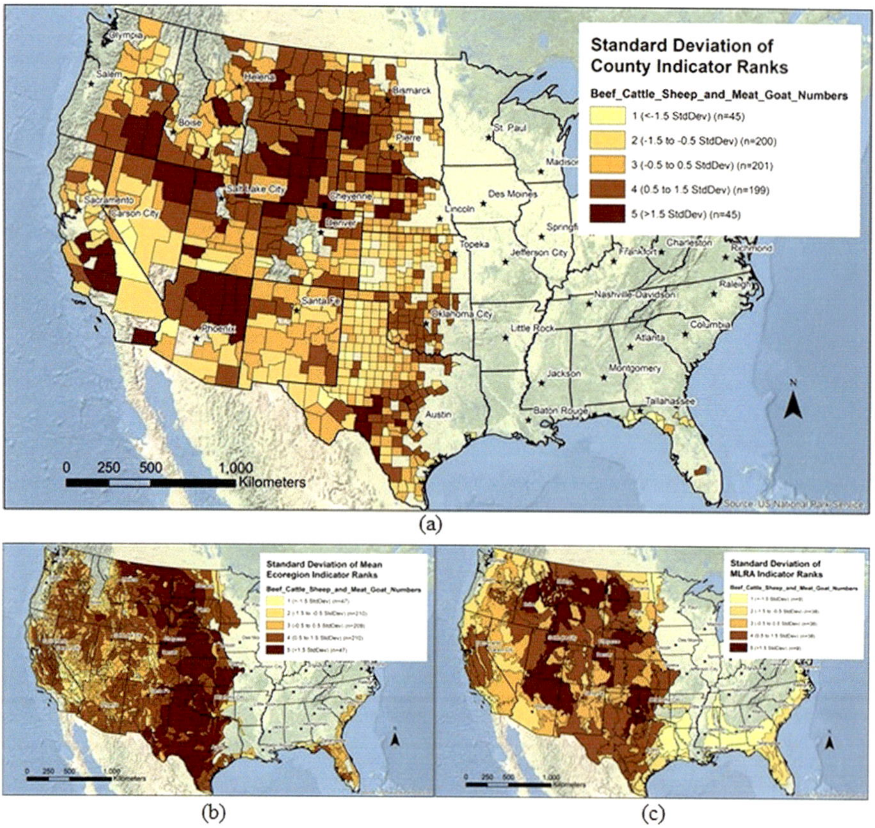

FIGURE 2.8
Number of domestic livestock on rangelands (beef cattle, sheep, and meat goats) by (a) County, (b) Ecoregion, and (c) MLRA.

assessment. Doing this on a percentage of total stream length by county may have been more informative. Despite this potential limitation, it is notable that Nemaha County in Kansas exhibits the greatest amount of impaired water, as evaluated here, with approximately 44 km of waterways in the data. In addition, 3 of the top 10 counties of the 690 evaluated in the assessment occur in Nebraska, including Gosper, Johnson, and Lincoln Counties supporting 18.4, 8.2, and 7.2 km, respectively. Collectively, the counties evaluated in the assessment exhibit about 1,150 km of impaired waters. The top five states ranked from the greatest to the least amount of impaired (length) are Texas, Kansas, Nebraska, South Dakota, and Montana with 245, 147, 125, 77, and 61 km, respectively.

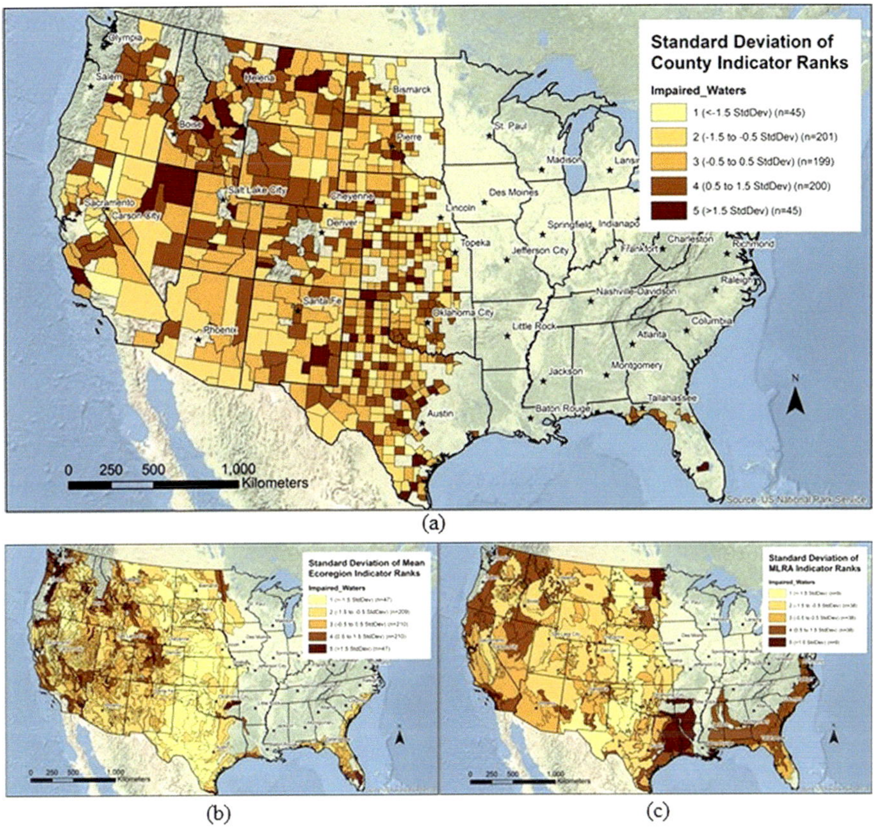

FIGURE 2.9
Percent of surface water on rangeland areas with significant deterioration of their chemical, physical, and biological properties, from acceptable levels (miles of impaired waters) by (a) County, (b) Ecoregion, and (c) MLRA

What Do the Ecological Indicators Mean by Themselves?

States and counties vary in how they rank against each other by indicator. Figure 2.10 shows the standardized rank average score for all ecological indicators summed together and standard deviations calculated. By adding the average rank for each county in a state, we can see how the states compare to each other (Figure 2.11). Oklahoma (540) scores the highest and Oregon (289) scores the lowest. Note that the maximum possible sum would be 900 if a state scored the highest on every indicator (recall that every indicator is normalized from 1 to 100 and there are 9 ecological indicators).

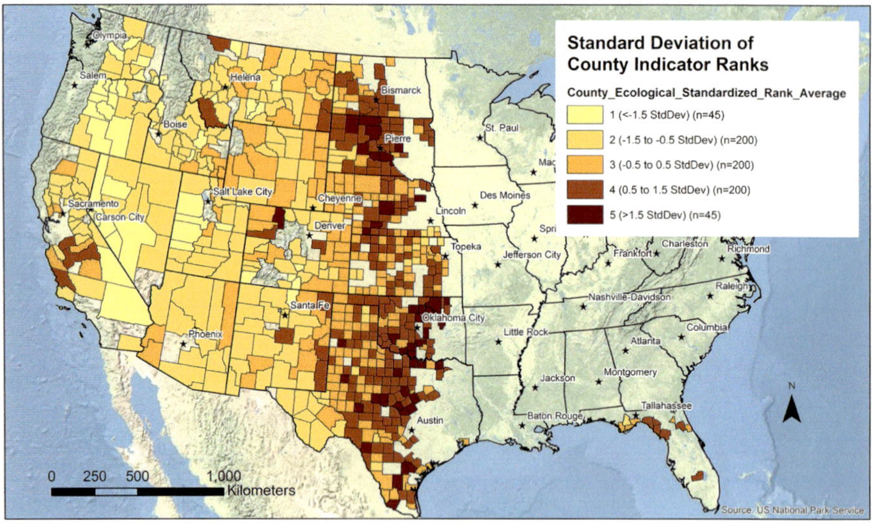

FIGURE 2.10
Distribution of standardized rank average scores for ecological indicators by standard deviation categories.

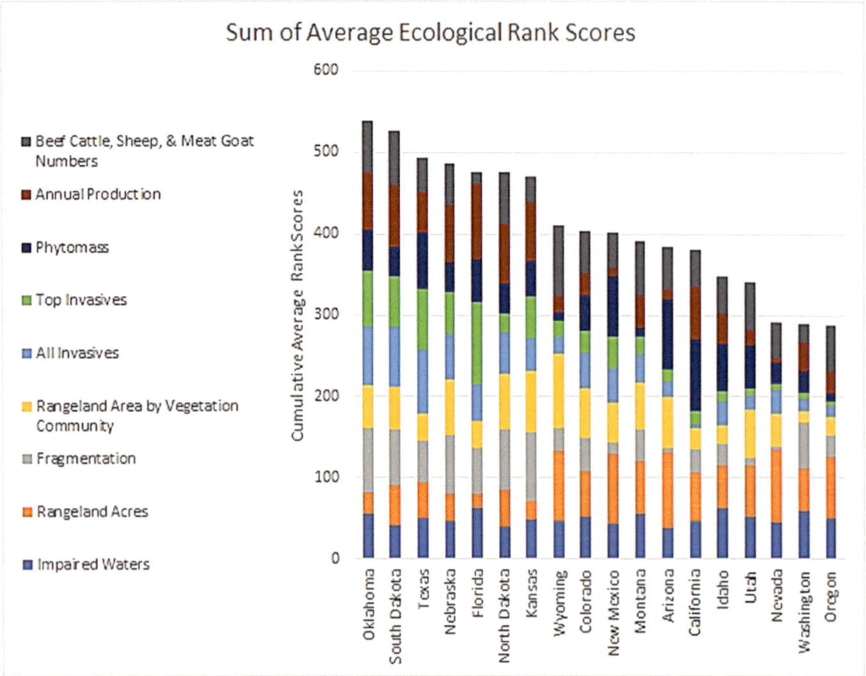

FIGURE 2.11
Sum of average ecological indicator standardized rank scores by state ranging from 289 to 540 out of 900.

If we examine counties within those two states, a clearer picture emerges. Figure 2.12 shows the Oklahoma rangeland counties with summed indicator scores ranging from 416 to 640. Figure 2.13 shows Oregon rangeland counties with summed indicator scores ranging from 216 to 379. In fact, no Oklahoma county scores are lower than the highest Oregon county (Klamath).

Does this mean that Oklahoma would be considered more ecologically sustainable compared to Oregon? Perhaps. What the data show is that Oklahoma ranks higher compared to all other rangeland nonurban counties. Whether it is better to rank in the highest quartile for *extent of land area in rangeland*, *rangeland area by vegetation community*, or *fragmentation of rangeland and rangeland plant communities* is up for discussion.

We can see how the rangeland counties are spatially distributed based upon their rank categories (Figure 2.10). It appears that counties in Great Plains (both northern and southern) regions are doing generally better than the rest of the counties from an ecological perspective, as indicated by the darker colors on the map. In contrast, counties in the Southern Great Basin and eastern Pacific Northwest are doing worse from an ecological perspective, as indicated by the lighter colors on the map (Figure 2.10). It is important to recall the suite of indicators used in the assessment, and perhaps even more critical to examine the data used to represent each indicator. Considering Oklahoma and Oregon again, if we examine the indicators used to derive Figure 2.10, some basic trends emerge. Counties in Oklahoma tend to have large numbers of cattle (Figure 2.8), high annual production (Figure 2.7), low numbers of invasive species (Figure 2.4), and very few of the especially problematic invasive species (Figure 2.5). In contrast, counties in Oregon exhibit a high degree of VDep (Figure 2.3) owing in large part to the presence of invasive annual grasses (Figure 2.4), high numbers of invasive species, moderate to low levels of productivity (Figure 2.7), and relatively low amounts of standing phytomass (Figure 2.6). While some counties are relatively intact, as indicated by low levels of fragmentation and high rangeland area, this is not enough to overcome the scores of the other indicators.

From this analysis, a few generalities can be justifiably retrieved. First, the sum of ranks tends to increase from west to east. Second, counties tend to have less land in the public domain farther to the east. Third, the sum of ranks and the relative position of each county within the ranking is heavily dependent on the size of the county. For example, larger counties have a greater likelihood of having significant waterways and therefore impaired water. The same is true for total range acres, counts of invasive species, presence of problematic invasive species, and livestock numbers. Indicators not dependent on county size include fragmentation, rangeland area by plant community, mean phytomass, and annual production. These issues can generate misleading or confounding conclusions. For example, consider Figure 2.11 where Wyoming exhibits a higher score for cattle numbers than Texas. This is counterintuitive since Texas has 3,840,000 cattle or 2.6 times the number in

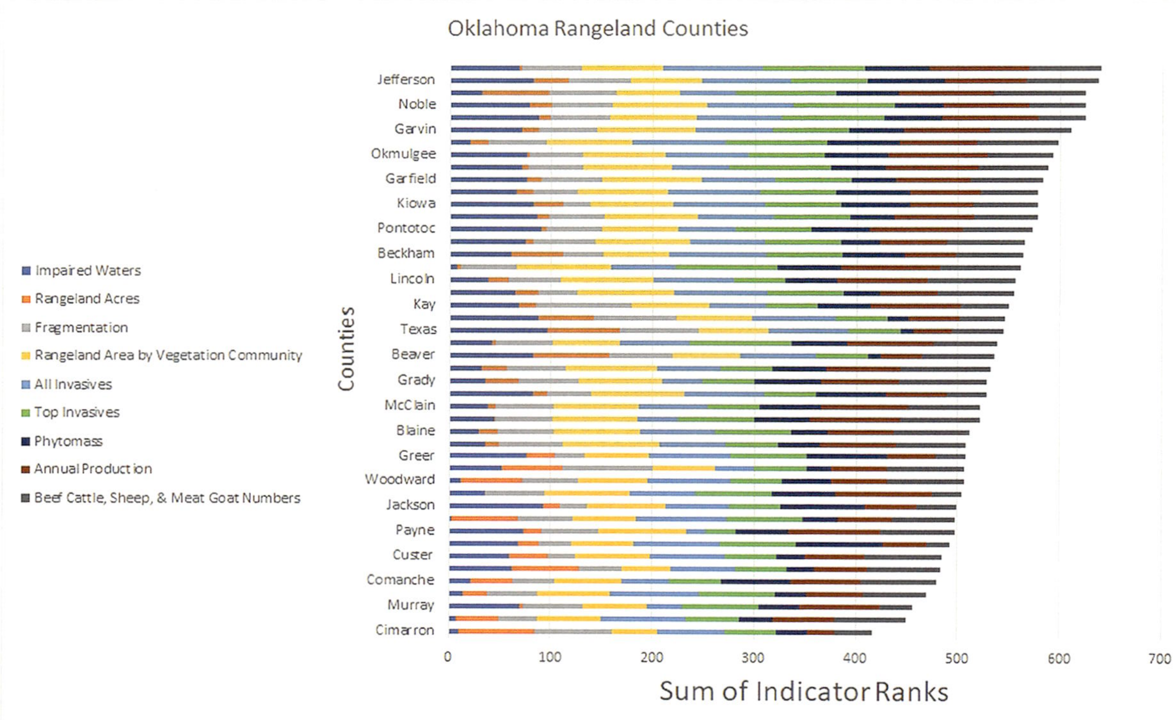

FIGURE 2.12
Oklahoma County ecological indicator standardized rank sums ranging from 416 to 640 out of 900.

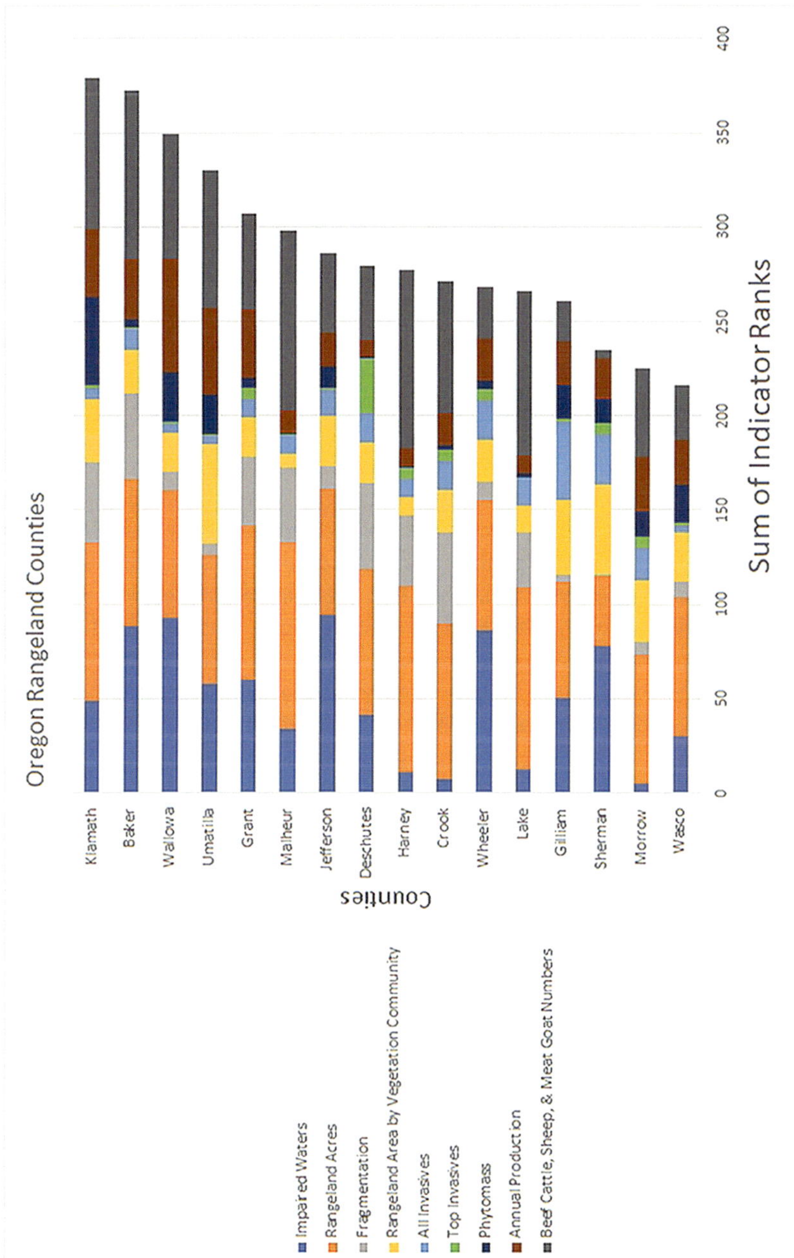

FIGURE 2.13
Oregon County ecological indicator standardized rank sums ranging from 216 to 379 out of 900.

Wyoming (as of 2017 in this assessment). So, while the sum of ranks for counties indicates higher scores, this is not the same as greater numbers of cattle overall at the state level. This issue suggests that further normalizing each area-dependent indicator may be prudent to facilitate more logical conclusions. However, this begs the question, should the total county area be used to normalize or should the amount of rangeland in each county be used? In many cases, the end result would likely be the same, but exceptions to this rule would surely occur.

In addition to the issue of area dependence, the results of the assessment and the individual scores are subject to interpretation, which is influenced by the nuances associated with each indicator. This is especially evident when considering the invasive species, mean phytomass, and VDep. Moreover, in the case of VDep, the results for this indicator are subject to our relative knowledge and understanding of contemporary vegetation conditions and how they compare to hypothesized distributions during pre-Eur-American settlement. Some vegetation types may be more highly studied and therefore more richly understood with respect to VDep. Thus, regions with more deeply studied vegetation types may have more appropriate depictions of VDep. For example, given the emphasis on and resources used for studying sagebrush ecosystems, it is possible that we have a greater understanding of how invasive species and changing fire regimes affect departure scores, compared with a relatively obscure ecological system such as Northwestern Great Plains Shrubland (Comer et al., 2003). The states with the lowest departure ratings (greater amounts of appropriate rangeland vegetation types and structure stages) include western Wyoming, northwestern South Dakota, western Kansas, and southwestern Arizona.

Lastly, the apparent lack of invasive species at the eastern and more southern extent of rangelands (Figures 2.4 and 2.5) may be due to a true lack of invasive species or to a relative lack of reported observations. Some private landowners may be reluctant to report the presence of invasive species or ecological conditions in general, so it is not possible to know with certainty what is causing the lack of invasive species in some counties.

Note

1 "Rangeland is defined by the NRI as a land cover/use category on which the climax or potential plant cover is composed principally of native grasses, grass-like plants, forbs, or shrubs suitable for grazing and browsing, and introduced forage species that are managed like rangeland. This includes areas where introduced hardy and persistent grasses, such as crested wheatgrass, are planted and such practices as deferred grazing, burning, chaining, and rotational grazing are

used, with little or no chemicals or fertilizer being applied. Grasslands, savannas, many wetlands, some deserts, and tundra are considered to be rangeland. Certain communities of low forbs and shrubs, such as mesquite, chaparral, mountain shrub, and pinyon-juniper, are also included as rangeland" (NRCS, 2021).

Literature Cited

Balch, J.K., Carroll, M.K., Bradley, B.A., 2012. The invasive grass-fire cycle in the U.S. Great Basin. *Teaching Issues and Experiments in Ecology* 8, 1–28.

Center for Invasive Species and Ecosystem Health, 2020. https://www.bugwood.org/index.cfm. University of Georgia (accessed 16 September 2020).

Comer, P., Faber-Langendoen, D., Evans, R., Gawler, S., Josse, C., Kittel, G., Menard, S., Pyne, M., Reid, M., Schulz, K., Snow, K., Teague, J., 2003. *Ecological systems of the United States: a working classification of U.S. terrestrial systems*. NatureServe, Arlington, Virginia.

Crooks, K.R., Burdett, C.L., Theobald, D.M., King, S.R.B., Di Marco, M., Rondinini, C., Boitani, L., 2017. Quantification of habitat fragmentation reveals extinction risk in terrestrial mammals. *Proceedings of the National Academies of Science* 114, 7635.

Coultrap, D.E., Fulgham, K.O., Lancaster, D.L., Gustafson, J., Lile, D.F., George, M.R., 2008. Relationships between western juniper (*Juniperus occidentalis*) and understory vegetation. *Invasive Plant Science and Management* 1, 3–11.

[EPA] Environmental Protection Agency, 2020. *303(d) impaired waters*. Waters Geospatial Downloads. https://www.epa.gov/waterdata/waters-geospatial-data-downloads (accessed 16 September 2020).

Miller, R., Svejcar, T., Rose, J.A. 2000. Impacts of western juniper on plant community composition and structure. *Journal of Range Management* 53, 574–585.

[NASS] National Agricultural Statistics Service, 2019. 2017 census of agriculture. https://www.nass.usda.gov/Publications/AgCensus/2017/index.php (accessed 20 April 2020).

[NRCS] Natural Resources Conservation Service, 2021. 2018 national resources inventory rangeland resources assessment. https://www.nrcs.usda.gov/wps/portal/nrcs/detail/national/technical/nra/nri/results/?cid=nrcseprd1343025 (accessed 25 January 2021).

Reeves, M.C., 2016. *Development of the rangeland vegetation simulator: a module of the forest vegetation simulator*. Joint Fire Science Final Report JFSP 14-S-01-01, 12-1-02-15, 12-1-02-15.

Reeves, M.C., Krebs, M.C., Leinwand, I., Theobald, D.M., Mitchell, J.M., 2018. *Rangelands on the edge: quantifying the modification, fragmentation, and future residential development of U.S. rangelands*. General Technical Report RMRS-GTR-382. US Department of Agriculture, Forest Service, Rocky Mountain Research Station, Fort Collins, CO, p. 36.

Reeves, M.C., Lankston, R., 2020. Rangeland production monitoring service: in season forage and fuel projections. US Department of Agriculture, Forest Service,

Rocky Mountain Research Station. https://www.fs.usda.gov/rmrs/tools/rangeland-production-monitoring-service-season-forage-and-fuel-projections (accessed 18 December 2020).

Reeves, M.C., Mitchell, J.E., 2011. Extent of coterminous US rangelands: quantifying implications of differing agency perspectives. *Rangeland Ecology and Management* 64, 16.

Rollins, M., 2009. LANDFIRE: a nationally consistent vegetation, wildland fire, and fuel assessment. *International Journal of Wildland Fire* 18(3), 235–249. doi:10.1071/WF08088

Theobald, D.M., 2013. A general model to quantify ecological integrity for landscape assessments and US application. *Landscape Ecology* 28(10), 1859–1874.

3

Social and Economic Rangeland Sustainability Indicators: The Challenge of Social Indicators and Sustainability

Social indicators are commonly used to assess where a society stands relative to questions such as quality of life, inequality, and environmental goals. Generally, social indicators are either objective or subjective. Objective indicators are statistical time series, often from official sources, that are significant for "any independent observer" (Land et al., 2012). Census data and unemployment percentage information are examples. Subjective indicators are derived from the point of view of individual assessments, such as self-reported health or survey data. Survey data for community satisfaction or sense of place are examples. The indicators used here are all objective indicators. This was necessitated by the spatial scale of this assessment.

RANK SCORING

For all social, economic, and ecological indicators, raw data for each indicator were ranked among the 690 rangeland counties. Counties are categorized by how many standard deviations they are from the mean of all counties. All analysis and discussion in this chapter are based on counties, ecoregions, MLRAs, or congressional districts (CD) compared to all of the other counties, ecoregions, MLRAs, or CD, respectively, using the standard deviation categories.

Rangeland sustainability more generally falls under the broader umbrella of sustainability in agriculture. Within the discussion of agriculture, the tools or approaches utilized in studies "often neglect to define social sustainability and lack stringent operationalization" (Janker and Mann, 2020, p. 1671). Using the SRR as a framework, a robust tool for evaluating rangeland sustainability has been developed (Fox et al., 2009). Within this tool, social and economic indicators were developed to assess the "economic and social capacity to maintain or change a social system" (Fox et al., 2009, p. 599).

Nonetheless, establishing an empirically valid relationship between ecological systems and social systems remains an elusive goal when applied

DOI: 10.1201/9781003279747-3

at this spatial scale. At best, we are identifying associations between social and ecological indicators. These associations provide context and display the societal forces that affect areas that we have characterized as rangelands. We are mapping data at different levels of aggregation rather than estimating associations. But we do not attempt or imply that any spatial level of social or economic organization used here has a causal relationship to ecological conditions. We can infer that landowners or users of both public and private lands affect the landscape and in aggregate might affect an ecological system. That inference is not supported without local data and certainly not at this spatial scale. However, this does not preclude the identification of associations between social and economic forces and ecological conditions. Thus, we are not measuring social and economic sustainability directly. Rather, we are using indicators to establish an understanding of the state of the social and economic systems associated spatially with rangelands for a national assessment.

Larger Social Forces and Rangeland Counties

Rangeland counties are not necessarily synonymous with rural areas. Many counties that qualify by our criteria for rangeland status include large urban centers, counties adjacent to urban areas, or counties more suburban in their settlement patterns. This situation is related to the larger problem of classifying rural areas of the United States (Isserman, 2005; Wunderlich, 2016). Wunderlich (2016, p. 2) observed, "The decline of the rural population in nonmetropolitan areas and the commingling with urban areas are clearly a challenge for rationalizing rural-urban classifications." Spatial, cultural, and social boundaries between urban and rural areas are "blurring" (Lichter and Ziliak, 2017, p. 14). These processes occur parallel to changes in ecological conditions characterizing rangeland. Our indicators are capturing some of the social and economic changes affecting the counties in this assessment. Slack and Jensen (2020) point to many trends in rural America to which some of our indicators are related, such as depopulation, aging, out-migration, growing ethnoracial diversity, and retirement and amenity-based migration.

These trends are ever-present in many rangeland counties and make large-scale assessments of sustainability particularly challenging. Assessments are difficult because the social sustainability of any place is a tangled web of social, demographic, economic, and cultural forces that can simultaneously be local, regional, national, and international in nature. As social forces work on rangeland areas at all spatial scales, we must be careful not to make specious links between the ecological conditions and social conditions. For example, some regions with a high percentage of rangeland are associated with high levels of poverty, while others with the same high percentage are associated with relative economic prosperity. This situation is not uncommon when evaluating the spatial distribution of poverty in rural America. The research literature that focuses on rural poverty is deep (Tickamyer et al., 2017). As described in Chapter 1, the SRR process confronted the difficulties inherent in the identification of social and economic indicators of rangeland sustainability while in the presence of larger, societal forces. With that in mind, we created a list of social and economic indicators from which a final set was chosen.

Indicator Data

The data used to populate the indicators has varied availability and usability at the regional and national scales. The original indicators were developed by the SRR Criteria and Indicator Process (SRR, 2003; Mitchell, 2010). The revised indicators are correlating indicators based primarily on available data sources and are used in this national assessment. In some cases, the data sources have changed from what was originally envisioned. As noted in Table 3.1, several of the indicators use data coming from the Census of Agriculture (NASS, 2019). These were collected on a county basis using both the Census of Agriculture publication and the Quick Stats and the Census Data Query Tool online tools provided by NASS (2020a). We had originally tried to get data only for those operators reporting that they had rangeland, but that was not possible given privacy and nondisclosure laws and regulations. Hence, those data are for all producers (crops and livestock, including livestock not on rangelands) within a county. Because we are just looking for indicators and not actual numbers, we do not view this as a significant impediment. The Census of Agriculture is conducted every five years, so data are included from 2017 when available. Some data have only been collected in the more recent Censuses of Agriculture (e.g., internet access). In most cases for the Census of Agriculture data, calculations were made to develop the values desired in the indicators. Most social indicator data used for this assessment is readily available from the US Census Bureau and the US Department of Commerce. These data are easily accessed at the county level and usually have time series available also. Table 3.1 shows the social and economic indicators used in this national assessment.

Because there are counties where NASS has a nondisclosure requirement, we needed to generate some of the values used for our assessment. Our solution was to find a state average from those counties that reported a value and then use that value for any county that had a nondisclosure. Counties with

TABLE 3.1

Economic and Social Indicators Used in the National Assessment

Original Indicator	Revised Indicator	Data Source
27. The value of forage harvested from rangeland by livestock	The value of forage harvested from rangeland by livestock	USDA National Agricultural Statistics Service, Census of Agriculture
32. Rate of return on investment for range livestock enterprises	Rate of return on investment for range livestock enterprises	USDA National Agricultural Statistics Service, Census of Agriculture
33. Area of rangelands under conservation ownership or control by conservation organizations	Area and number of operations with conservation easements	USDA National Agricultural Statistics Service, Census of Agriculture
38. Income inequality	Income inequality—Gini coefficient	US Census, American Community Survey
39. Index of social structure quality	DCI	EIG
46. Years of education	1. No high school diploma: Percent of the 25+ population without a high school diploma or equivalent	US Census, American Community Survey
	2. Housing vacancy rate: percent of habitable housing that is unoccupied, excluding properties that are for seasonal, recreational, or occasional use	US Census, American Community Survey
	3. Adults not working: percent of the prime-age population (25–64) not currently in work	US Census, American Community Survey
36. Poverty range (general)	4. Poverty rate: percent of the population living under the poverty line	US Census, American Community Survey
	5. Median income ratio: median household income as a percent of the state's median household income	US Census, American Community Survey
	6. Change in employment: percent change in the number of jobs	US Census, County Business Patterns
41. Federal transfers by categories (individual, infrastructure, agriculture, etc.)	Federal, state, and local transfers to ranches	USDA National Agricultural Statistics Service, Census of Agriculture
	Per capita total federal transfers	

(Continued)

TABLE 3.1 (Continued)

Economic and Social Indicators Used in the National Assessment

Original Indicator	Revised Indicator	Data Source
43. Sources of income and level of dependence on livestock production for household income	Percent of farm income from recreation	USDA National Agricultural Statistics Service, Census of Agriculture
	Level of dependence on livestock production for household income	USDA National Agricultural Statistics Service, Census of Agriculture
44. Employment diversity	Employment diversity	US Department of Labor, Bureau of Labor Statistics
45. Agriculture (ranch/ farm) structure	Gross agricultural sales, livestock per farm, livestock sales, acres in production, and business organization	USDA National Agricultural Statistics Service, Census of Agriculture
	Number of beef cows, sheep, and meat goats per farm (type of commodity produced)	USDA National Agricultural Statistics Service, Census of Agriculture
	Percent of farms in the county classified as having beef cattle	USDA National Agricultural Statistics Service, Census of Agriculture
	Livestock numbers sold	USDA National Agricultural Statistics Service, Census of Agriculture
	Acres of rangeland and pastureland	USDA National Agricultural Statistics Service, Census of Agriculture
	Percent of family farms (individual, partnership, family corporation, and limited liability corporation business organizations)	USDA National Agricultural Statistics Service, Census of Agriculture
47. Value produced by agriculture and recreation as percent of total economic output	Value produced by agriculture and recreation as percent of total economic output	US Department of Commerce, Bureau of Economic Analysis
48. Unemployment rate	Percent unemployed	US Department of Labor, Bureau of Labor Statistics

(Continued)

TABLE 3.1 (Continued)

Economic and Social Indicators Used in the National Assessment

Original Indicator	Revised Indicator	Data Source
49. Land tenure, land use, and ownership patterns by land size classes	Land tenure	USDA National Agricultural Statistics Service, Census of Agriculture
	Land use	
	Ownership patterns	
50. Population pyramid and population change	Population change, 2013 to 2017	US Census, American Community Survey
	Percent population age 65+	US Census, American Community Survey
	Percent population age 18 or younger	US Census, American Community Survey
53. Income by work location versus residence	County earnings adjusted for residence	US Department of Commerce, Bureau of Economic Analysis

no data as opposed to nondisclosure were kept at zero for the assessment. A similar method was used for social data. This method was used by the Economic Innovation Group to account for a small number of counties lacking sufficient data to calculate a Distressed Communities Index (DCI), so we applied the same method to other datasets.

Indicator data for the 690 counties were ranked from low (1) to high (690). Ranking was selected to compare rangeland county values. In the cases where there were multiple data sources for a single indicator, the rankings were averaged to come up with a ranking score for that indicator. The final rankings were then standardized from 0 to 100 using the formula

$$(\text{indicator score} - \text{rank minimum})/(\text{rank maximum} - \text{rank minimum}) * 100.$$

That standard score was then used to represent each county for mapping and further analysis. For the assessment of sustainability, standard scores for economic and social indicators were averaged and standardized so that there would be a single standard score for economic indicators and social indicators. These were used in conjunction with the ecological indicators to assess sustainability.

We further analyzed the data by major land resource area (MLRA), Küchler Ecoregions (KE), and congressional districts (CD). For each of these analyses, the percent of each county in an MLRA, KE, or CD was calculated using ArcMap (ESRI, 2019). The county-level data were then allocated to each unit and summed for that unit. Once that was done, the standardization process

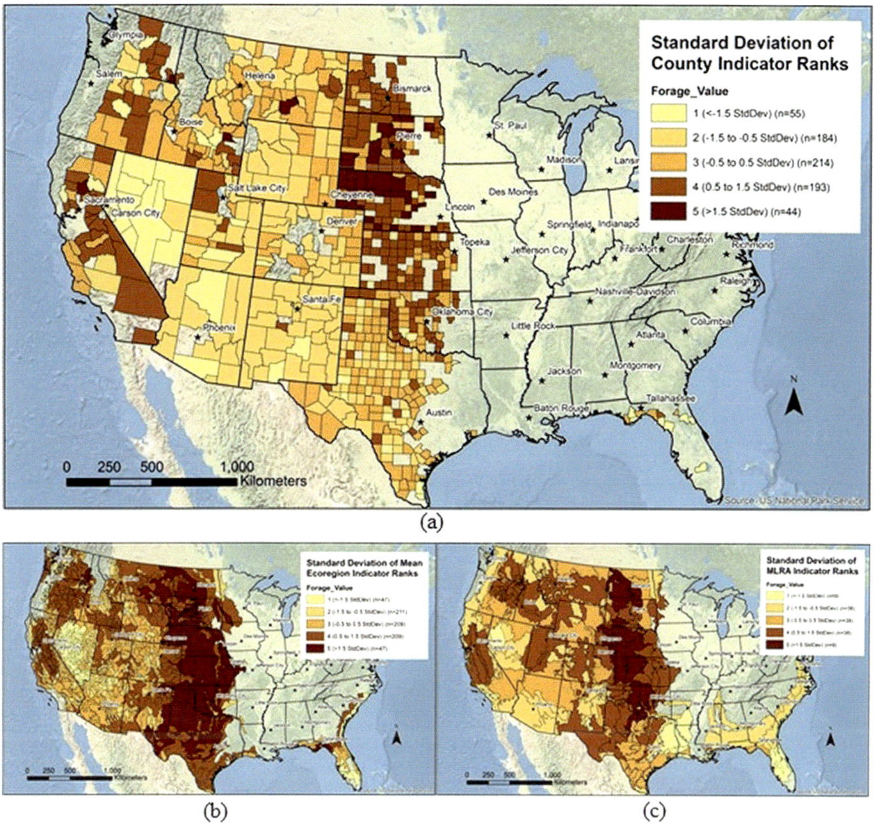

FIGURE 3.1
Forage value by (a) County, (b) Ecoregion, and (c) MLRA.

used for counties was completed both at the MLRA, KE, or CD unit level for each indicator and for the economic and social indicator levels for use in this national assessment.

The Value of Forage Harvested from Rangeland by Livestock

The *value of forage harvested from rangeland by livestock* was selected as an indicator of an ecosystem service (Figure 3.1). Forage is vegetation that is consumed by cattle, sheep, or goats during the grazing season. As such, it is an intermediate product in the livestock life cycle. There are several ways that value has been placed on forage. These values are reported by the NASS (2020b) from an annual survey of cash rents on a county basis. The cash rent for pasture was selected as the indicator for the value of forage for this assessment.

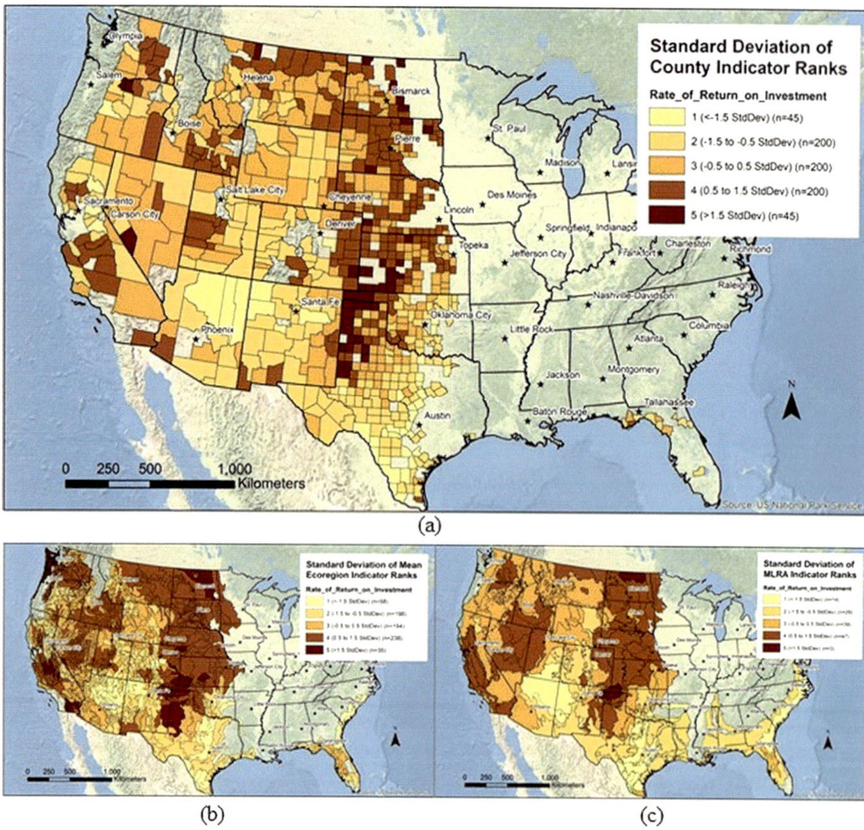

FIGURE 3.2
Rate of Return on Investment in Range Livestock Ranches by (a) County, (b) Ecoregion, and (c) MLRA.

Rate of Return on Investment for Range Livestock Enterprises

Data to calculate the *rate of return on investment for range livestock enterprises* was derived from the Census of Agriculture (NASS, 2019). This indicator shows the financial health of ranch enterprises (Figure 3.2). Since we could not extract data directly for rangeland-based operations, it represents all farm and ranch operations in a county. We used net cash farm income by operation and divided that by the agricultural land and buildings, assets/operation to get the rate of return on investment.

Area and Number of Operations with Conservation Easements

While the original indicator sought the area under *conservation easements* by conservation organizations, there is no known source of such data for rangelands.

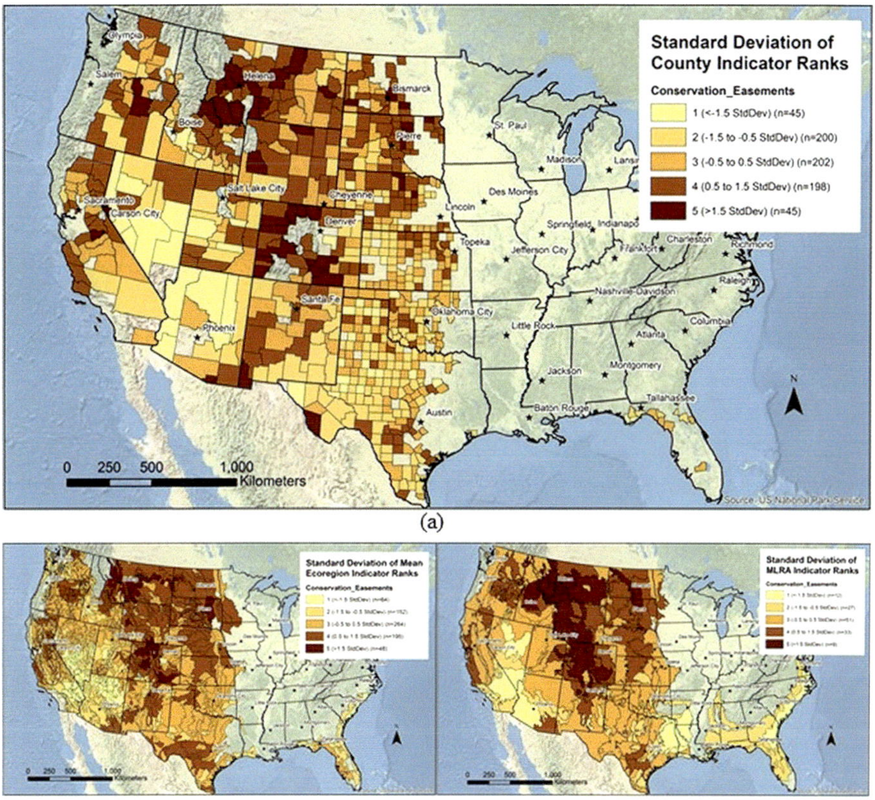

(b) (c)

FIGURE 3.3
Area and Number of Operations with Conservation Easements by (a) County, (b) Ecoregion, and (c) MLRA.

The Census of Agriculture (NASS, 2019) began collecting data on land under conservation easement by farms and acres (Figure 3.3) in 2017, without regard for which organization held the easement. While this only results in one time period for data, it is a start for measuring what we originally intended.

Income Inequality

This indicator (Figure 3.4) provides a measure of general *income inequality* across a community as it pertains to economic equity and stratification. Though numerous single value measures of income equality are available, we concluded that ready access and ease of interpretation made the Gini coefficient a logical choice (US Census Bureau, 2020a). We expect that lower values of this coefficient are associated with lower cohesion and social

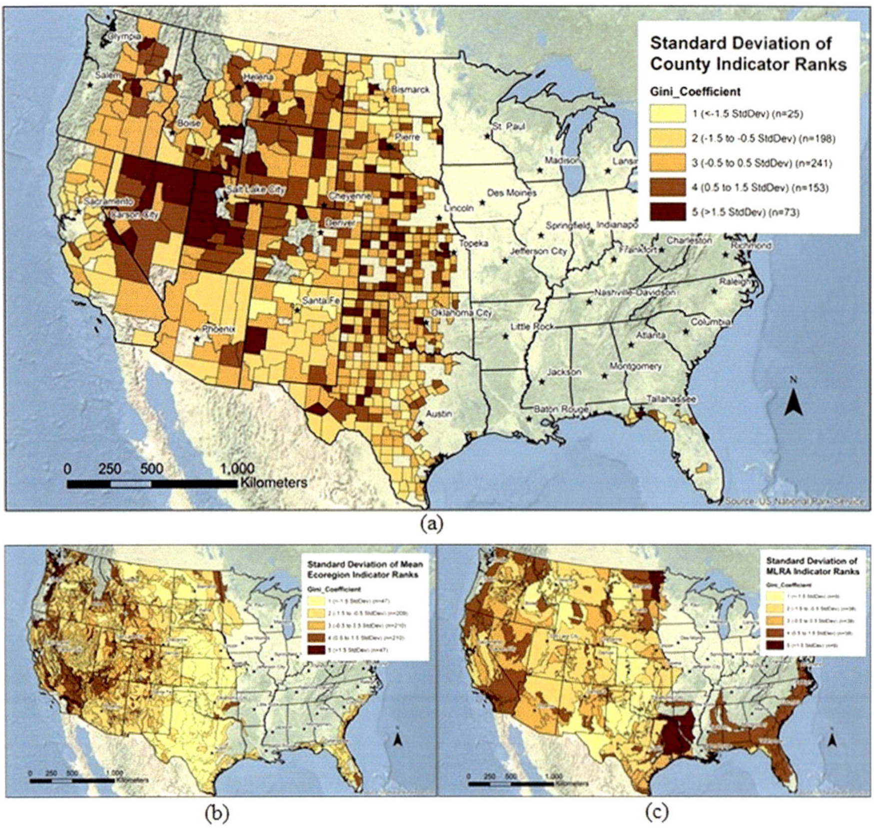

FIGURE 3.4
Income inequality (Gini Coefficient) by (a) County, (b) Ecoregion, and (c) MLRA.

integration in a county (Tanaka et al., 2009, p. 14). Gini coefficients are calculated on a scale from 0.0 (perfectly equal distribution of income) to 100.0 (one household has all the income). This allows us to use the coefficients without additional standardization.

For rangeland counties in 2017, the Gini coefficient ranged from 0.34 to 0.59 on a scale from 0.0 to 1.00. Figure 3.4 reflects a ranking where a lower coefficient is "better" than a higher coefficient based on deviation from the mean for all rangeland counties. The counties with darker shading have more equal distribution of income than those with lighter shading.

Distressed Communities Index

The original first approximation of the social and economic indicators discussed an "Index of Social Structure Quality" (Tanaka et al., 2009, p. 14).

This indicator was intended to offer a multidimensional measure of "the capacity of communities to address their quality-of-life and may indicate their willingness to address environmental issues" (Tanaka et al., 2009, p. 14). To create this indicator, we are using the *DCI* (Figure 3.5) from the Economic Innovation Group (EIG; Economic Innovation Group, 2018). The index also captures Indicator 36: poverty. EIG's multidimensional index is constructed of the following sub-indicators:

1. No high school diploma: percent of the 25+ population without a high school diploma or equivalent
2. Housing vacancy rate: percent of habitable housing that is unoccupied, excluding properties that are for seasonal, recreational, or occasional use

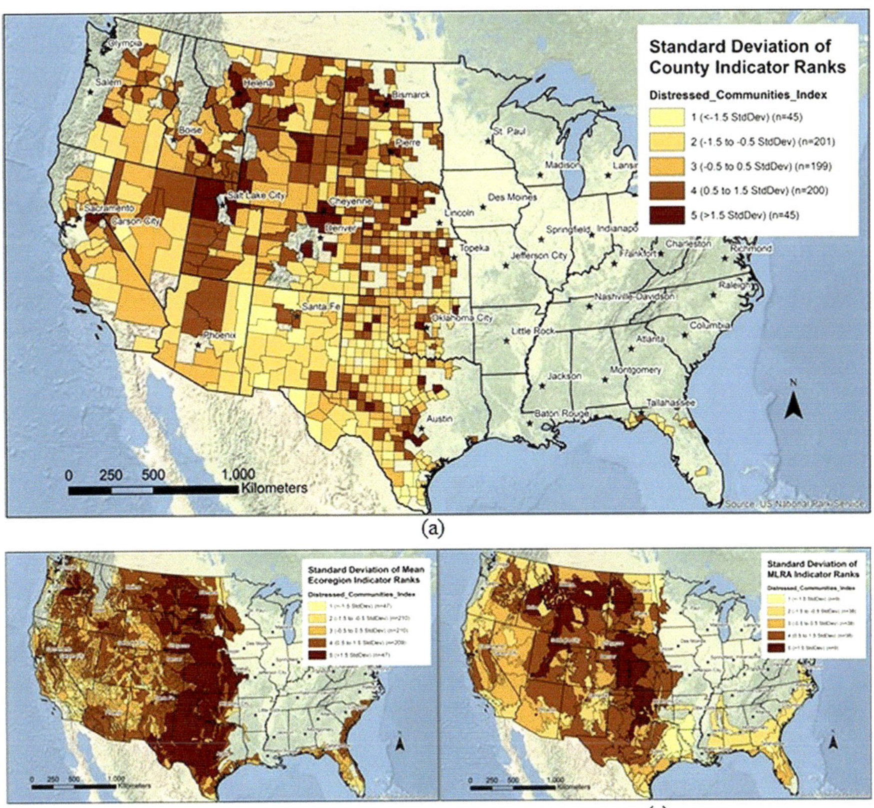

FIGURE 3.5
Distressed community index by (a) County, (b) Ecoregion, and (c) MLRA.

3. Adults not working: percent of the prime-age population (25–64) not currently in work
4. Poverty rate: percent of the population living under the poverty line
5. Median income ratio: median household income as a percent of the state's median household income (to account for cost-of-living differences across states)
6. Change in employment: percent change in the number of jobs
7. Change in establishments: percent change in the number of business establishments

The data for these indicators are drawn from American Community Survey five-year estimates and County Business Patterns (US Census Bureau, 2020a; US Census Bureau, 2020b). Data from counties is weighted equally, then ranked for each indicator. These ranks are then averaged and standardized to a range from 0.0 to 100.0.

EIG's original research then broke the ranks into quintiles. Communities in the lowest quintile were labeled "distressed" and those in the highest quintile were labeled "prosperous" (Economic Innovation Group, 2018, p. 3). Taken as a whole, this index measures the ability of a community to respond to economic and social change and the likelihood that it will possess or create sufficient social integration to mitigate "undesirable outcomes like reduced health, human capital, social capital" (Tanaka et al., 2009, p. 13). The counties with darker shading are doing better than those with lighter shading.

Regions with population growth appear to lean toward prosperous, such as the Wasatch Front, Denver, and Phoenix. The rural southwest has numerous distressed counties, as reflected in other indicators such as unemployment and lower population growth. Two regions of note are, again, western North Dakota and the Permian Basin area of west Texas; both were experiencing economic growth in 2017, and this is reflected in Figure 3.5.

Federal, State, and Local Transfers to Ranches

Originally this indicator was only federal transfers by category of transfer. While that is still a good indicator, in this case, the Census of Agriculture (NASS, 2019) has *federal, state, and local transfers to farms and ranches* (Figure 3.6). Hence, the economic indicator was renamed and focuses only on the category of agriculture. Data come from the amount from state and local government agricultural program payments and the amount from federal government program payments by county.

As a social indicator (Figure 3.7), *per capita transfer payments to individuals* can indicate the level of dependence a county has on many types of federal dollars such as Social Security, income maintenance, and farm payments. This indicator could be broken down into payments by type, such as payments

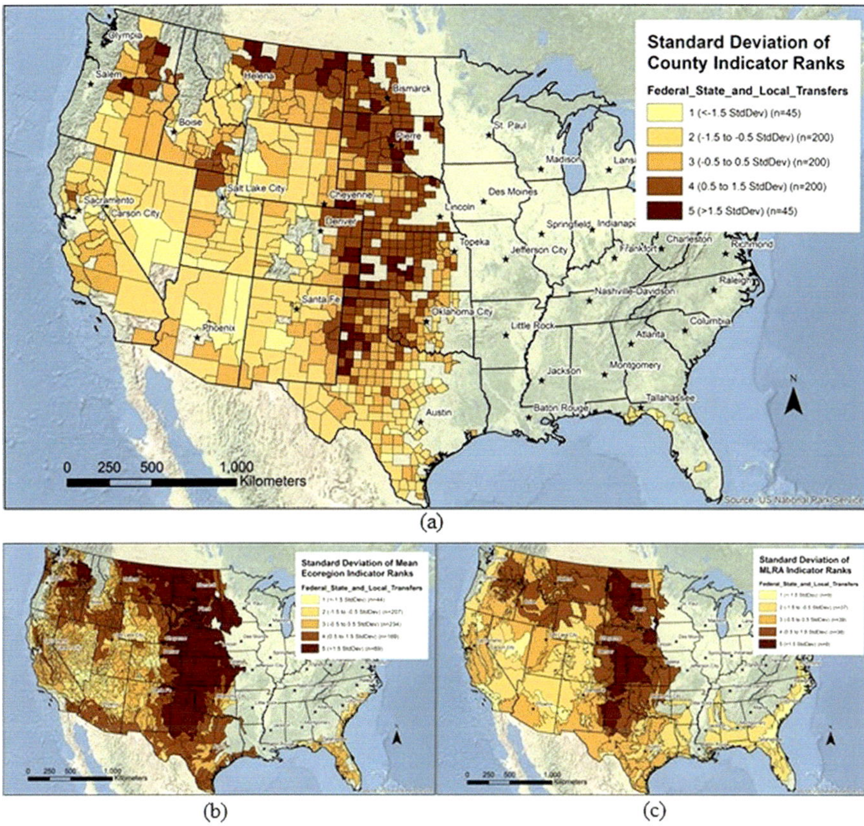

FIGURE 3.6
Federal, state, and local transfers to ranchers by (a) County, (b) Ecoregion, and (c) MLRA.

to farms, income support payments, and retirement-related payments. With that in mind, per capita total transfers to counties were used as a social indicator of a general dependence on federal payments.

Per capita total federal transfer payments by county show a spatial pattern that generally follows crop farming, poverty, aging, and retirement. Much of New Mexico is economically disadvantaged and also has a significant retirement component. This is reflected in Figure 3.7 by their total transfer ranking being above the mean for the group. Regions well-known as retirement destinations, such as Phoenix and central Oregon, are also ranked well above the mean. Areas experiencing growth like the Wasatch Front of Utah are receiving fewer transfer payments relative to the other rangeland counties. This can be seen by their rankings below the mean. Areas normally associated with crop farming in the Midwest also score above the mean for this indicator.

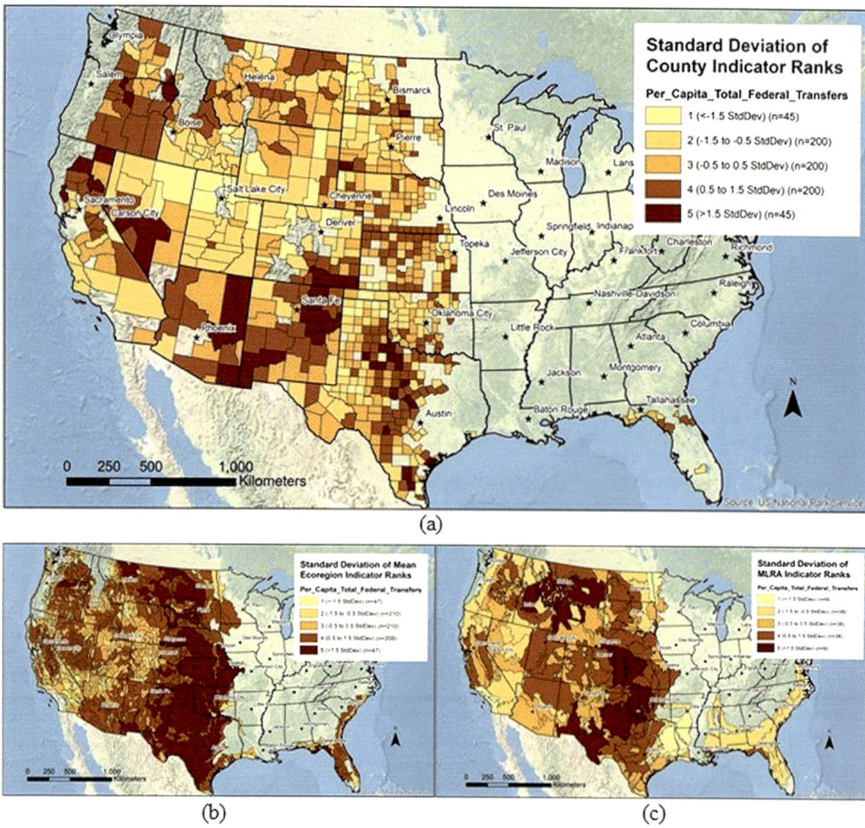

FIGURE 3.7
Per capita total federal transfer payments by (a) County, (b) Ecoregion, and (c) MLRA.

Sources of On-Farm Income and Level of Dependence on Livestock Production for Household Income

Data on *sources of on-farm income and level of dependence on livestock production for household income* were obtained on a county-level basis from the Census of Agriculture (NASS, 2019). These indicators are intended to be a measure of the degree of dependence that ranch families have on recreation and livestock production for household income (Figures 3.8 and 3.9). Higher dependence may make the ranch family more susceptible to things like droughts, low livestock prices, and their ability to follow sustainable grazing practices that may have a negative effect on ecological sustainability. This indicator was separated into two sub-indicators for analysis: 43A is the *percent of income from recreation* and 43B is the *level of dependence on livestock production for household*

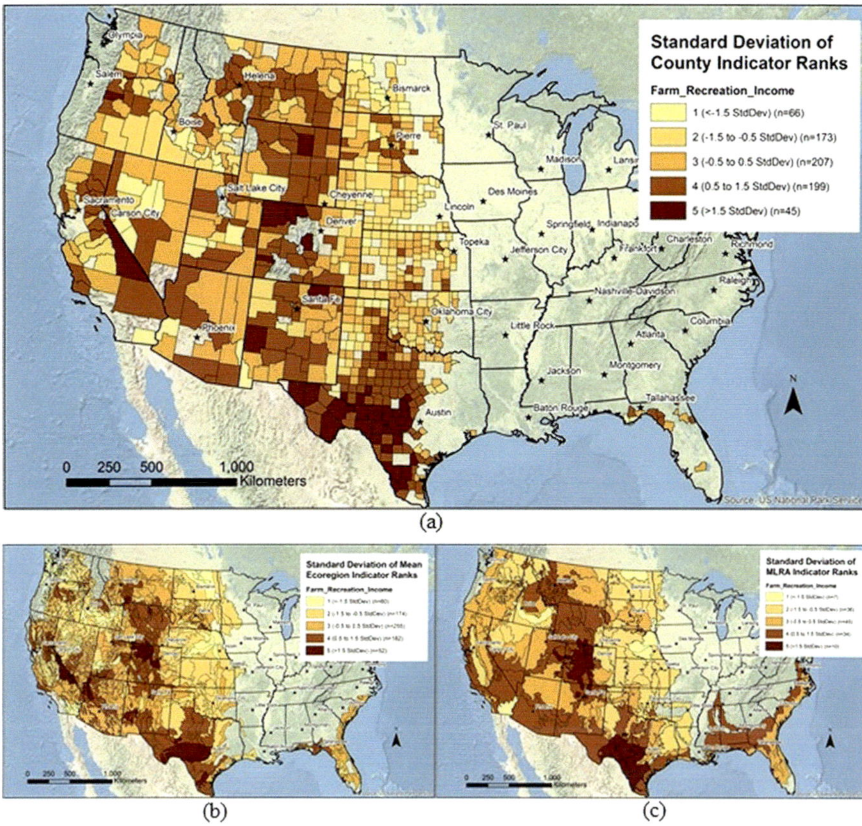

FIGURE 3.8
Percent of income from recreation by (a) County, (b) Ecoregion, and (c) MLRA.

income. While there are other sources of income on the ranch, we separated out recreation income as its own indicator. It was originally included as part of the level of dependence on livestock production. The level of dependence on livestock production was calculated as sales from beef cattle, sheep, wool, and meat goats as a percentage of total income from farm-related sources. This indicator likely overstates the dependence for household income, as we only used farm-related sources. Both indicators are susceptible to general economic fluctuations, as well as ecological changes on the land.

Employment Diversity

Data on *employment diversity* were obtained from the Bureau of Labor Statistics (BLS, 2020) on jobs by county. Because of data limitations at the

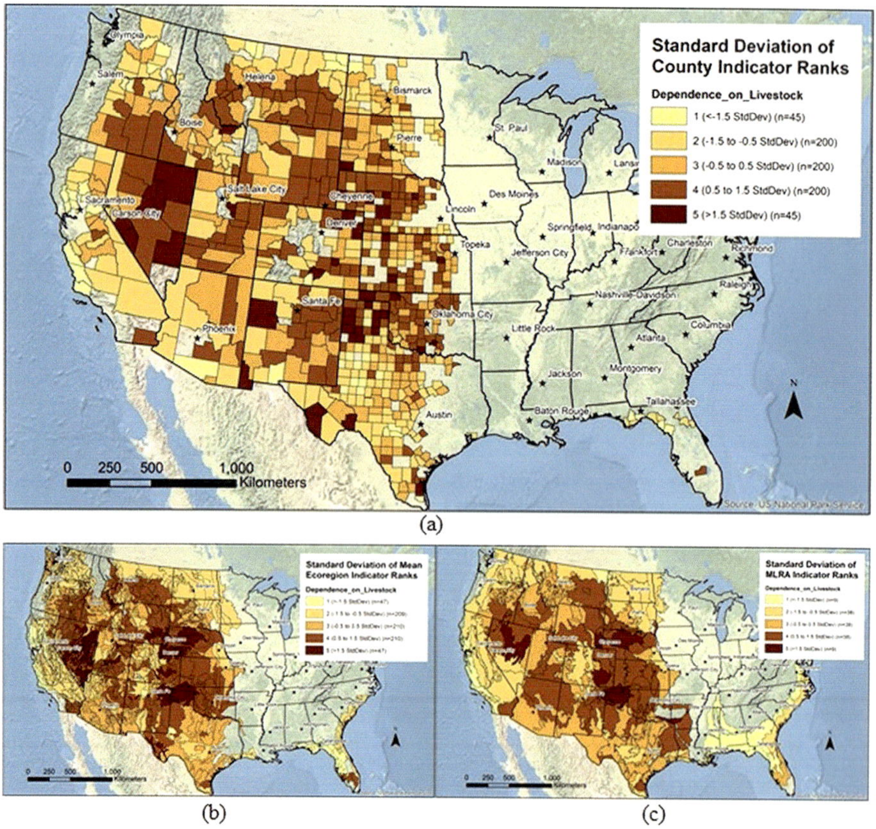

FIGURE 3.9
Level of dependence on livestock production for household income by (a) County, (b) Ecoregion, and (c) MLRA.

county level, aggregate data for federal, state, and local governments, natural resources and mining, construction, manufacturing, trade, transportation, and utilities, information, financial activities, professional and business services, education, and health services, leisure and hospitality, and other services were collected. The formula for Simpson's Diversity Index (SDI) was used to calculate an employment diversity score (Figure 3.10). SDI was originally created to measure species diversity in ecology. Its interpretation would be the same for counties. Calculations of SDI result in scores that are closer to 1 being more diverse economies and those closer to 0 being less diverse. We presume that a county economy that is more economically diverse is more resilient and has a better ability to adapt and respond to changes in conditions.

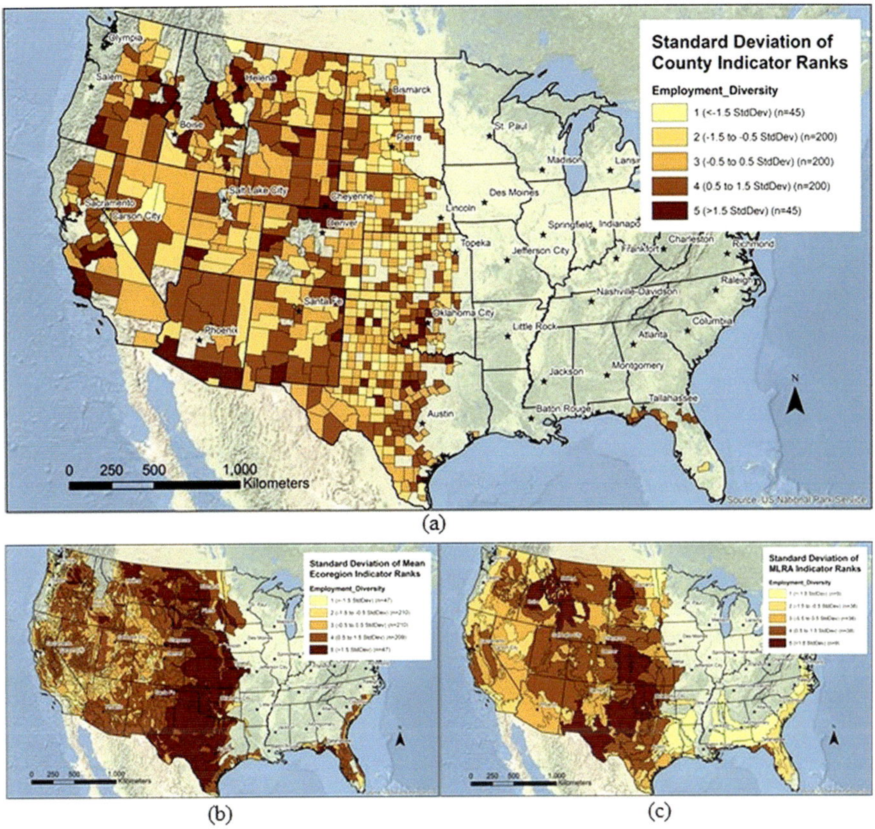

FIGURE 3.10
Employment diversity by (a) County, (b) Ecoregion, and (c) MLRA.

Agriculture (Ranch/Farm) Structure

Data on *agriculture (ranch/farm) structure* were obtained on a county-level basis from the Census of Agriculture (NASS, 2019). This is a multicomponent indicator made up of several sub-indicators. Data were collected on gross agricultural sales, type of farm (beef, sheep, goat), type of livestock sales, acres in private pasture and rangeland, and type of business organization. Each sub-indicator was ranked, and those were averaged to get a final rank for the indicator. This is an indirect indicator of the production capacity for food and fiber.

This indicator was divided into six parts. *Gross agricultural sales* are shown in Figure 3.11; the *number of beef cows, sheep, and meat goats per farm* (type of commodity produced) is shown in Figure 3.12; the *percent of farms in the county classified as having beef cattle* is shown in Figure 3.13; *livestock numbers*

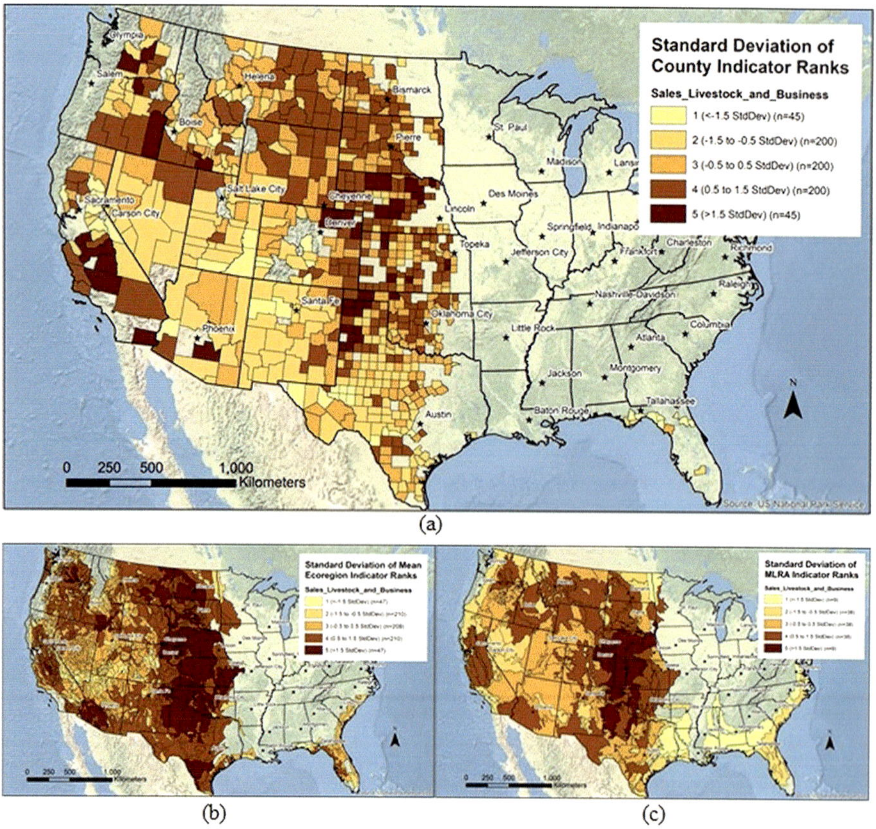

FIGURE 3.11
Gross agricultural sales by (a) County, (b) Ecoregion, and (c) MLRA.

sold is shown in Figure 3.14; the *acres of rangeland and pastureland* is shown in Figure 3.15; and the *percent of family farms* (individual, partnership, family corporation, and limited liability corporation business organizations) is shown in Figure 3.16.

Value Produced by Agriculture and Recreation Industries as Percent of Total Economic Output

Data on the *value produced by agriculture and recreation industries as percent of total economic output* were obtained on a county-level basis from the Bureau of Economic Analysis (BEA, 2020) for total gross domestic product and for agriculture, forestry, fishing, and hunting. This is the proxy for agriculture

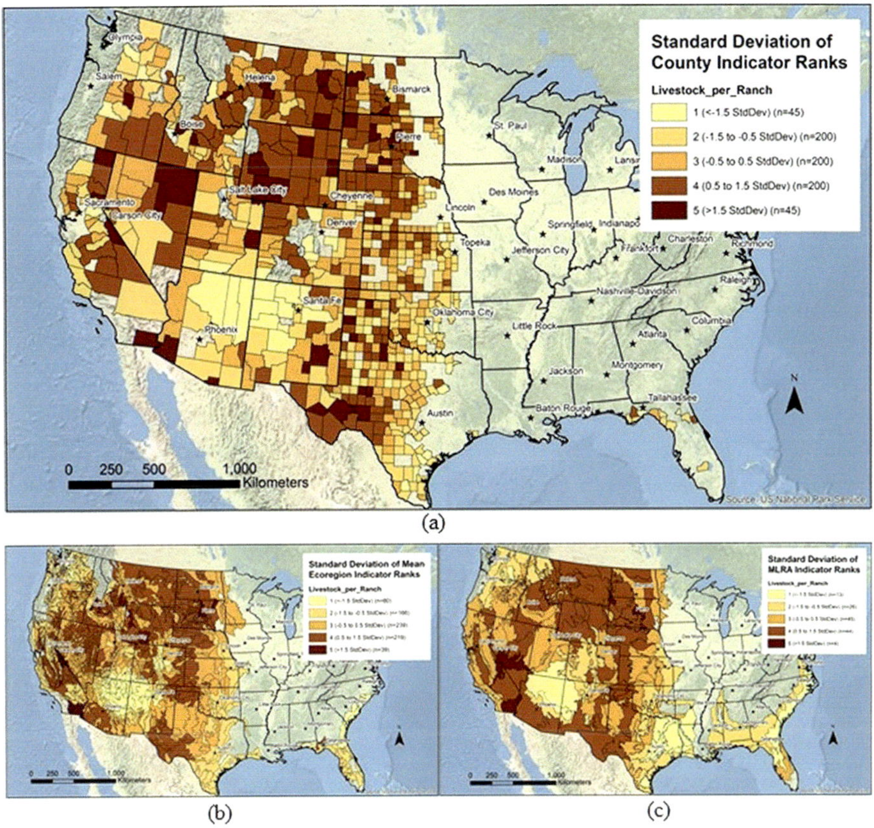

FIGURE 3.12
Number of beef cows, sheep, and meat goats per farm by (a) County, (b) Ecoregion, and (c) MLRA.

and recreation. This indicator includes the economic value produced from wildlands (rangeland and forestland) in our rangeland counties (Figure 3.17). Tracking these values is indicative of the pressures placed on rangelands. Nonmarket values are not included in the indicator, only those values that flow through the local economy.

Unemployment

Cross-sectional and time-series data for *unemployment* is readily available for counties. The original indicator included underemployment and discouraged workers data. Data for those parts of our original indicator can be obtained at the state level (US Census Bureau, 2020b). We felt that the county

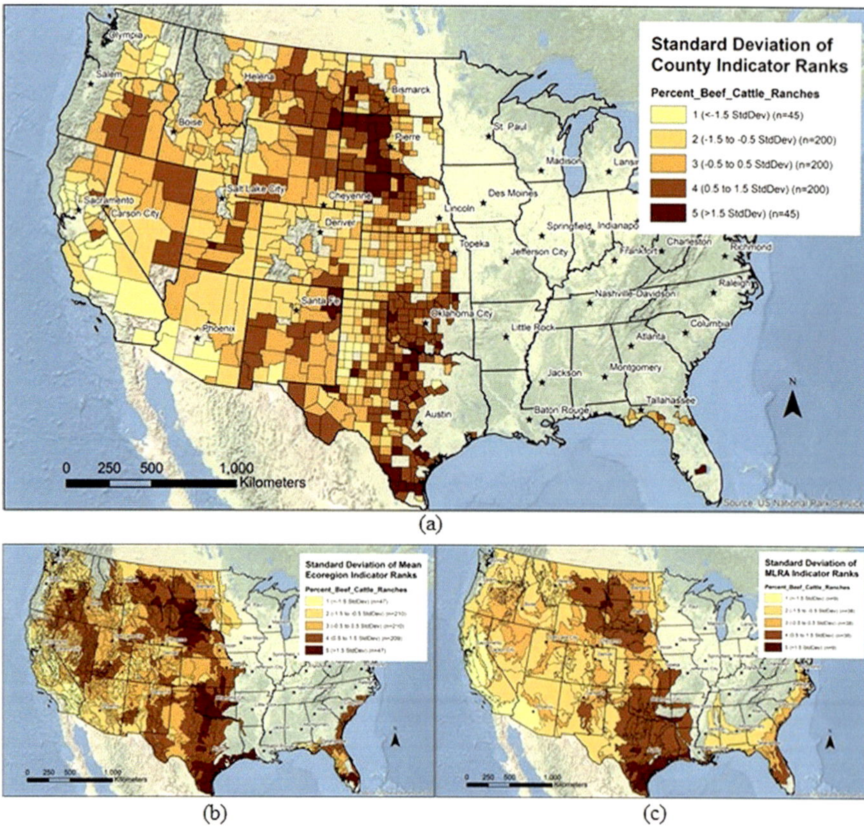

FIGURE 3.13
Percent of farms in the county classified as having beef cattle by (a) County, (b) Ecoregion, and (c) MLRA.

unemployment rate is a commonly used indicator that is readily understood by end users of this analysis (Figure 3.18).

Unemployment rates by county are expressed as a percentage of the county workforce who are available for work, are without work, and are seeking work. Part of this indicator's utility is that this number is familiar due to its use in media reports, etc. Unemployment rates were standardized to a range of 0 to 100 and ranked for each county for the analysis.

The county unemployment rates for rangeland counties presented in Figure 3.18 reflect our ranking approach where lower unemployment rates are "better" than higher. Thus, counties above the mean display low unemployment. We see two divergent possible trends. First, many areas

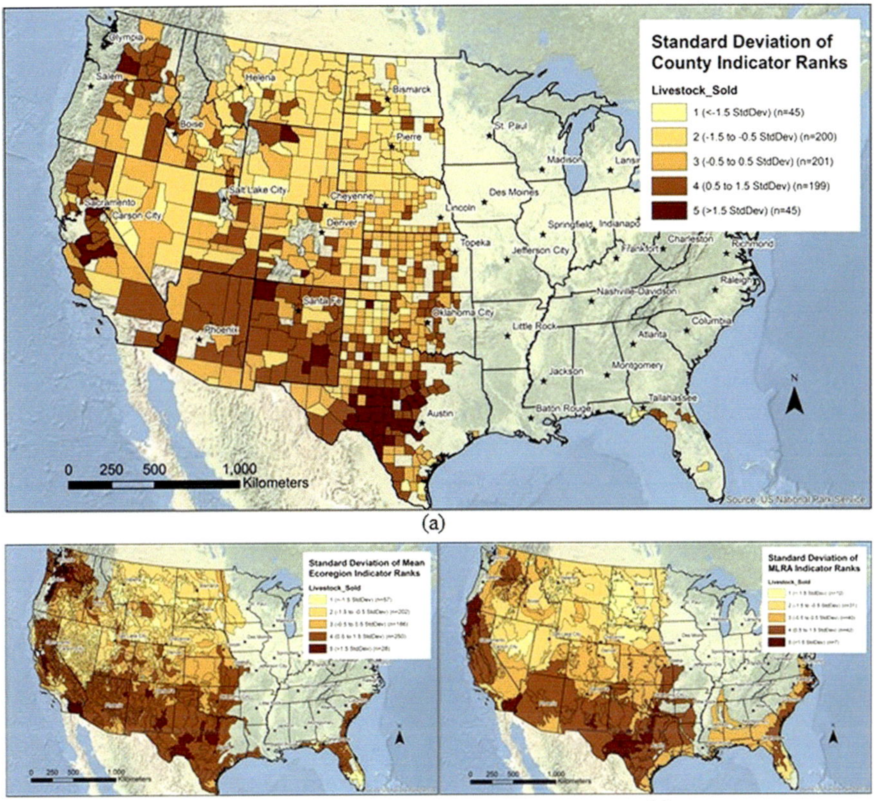

(a)

(b) (c)

FIGURE 3.14
Livestock numbers sold by (a) County, (b) Ecoregion, and (c) MLRA.

experiencing high growth have relatively low unemployment. Growth is creating jobs fast enough that anyone who wants a job can likely find one. Second, little growth has resulted in a loss of the working-age population, where those remaining have jobs and others have simply migrated out of the area.

Land Tenure, Land Use, and Ownership Patterns by Land Size Classes

The indicator *land tenure, land use, and ownership patterns by land size classes* is a conglomerate of many measures derived from the Census of Agriculture (NASS, 2019). In general, it measures changes in ownership, ownership

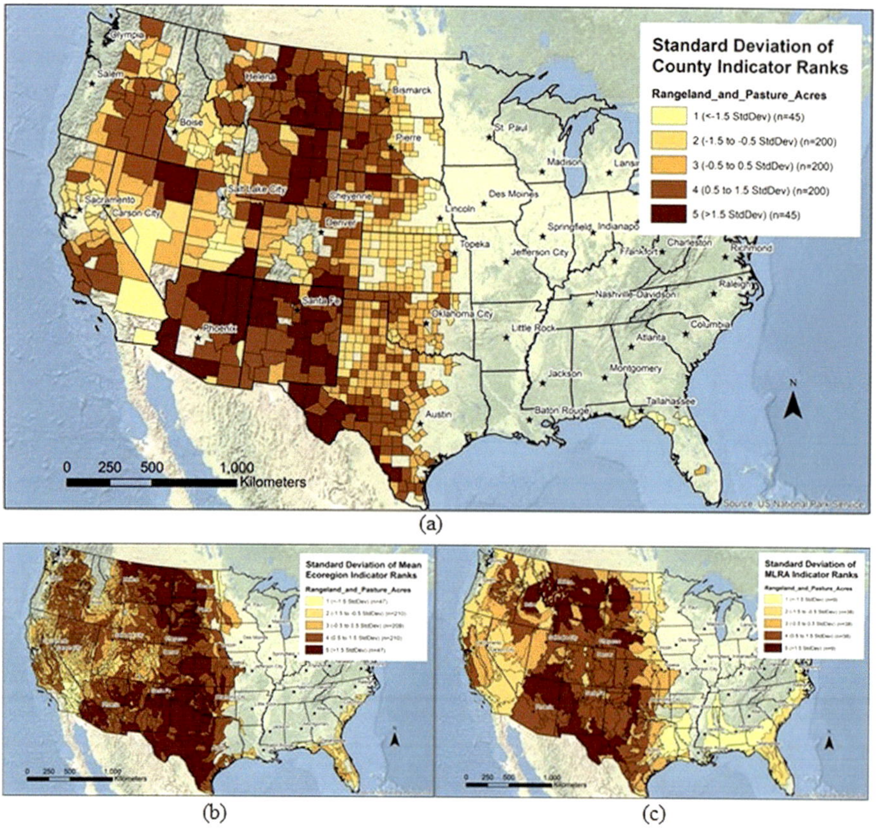

FIGURE 3.15
Acres of rangeland and pastureland in Ranches by (a) County, (b) Ecoregion, and (c) MLRA.

stability, and how the land is being used. Various sub-indicators were averaged to estimate a single value for the indicator. We were able to determine the percentage of farms in each tenure class (full time, part time, and tenant), the percentage of rangeland used, operator gender, principal occupation, residence, days worked off-farm, years on current or any farm, age, and internet usage. The ranks for each of these categories were averaged to get a final rank for the indicator.

This indicator was divided into three sub-indicators. *Land tenure* is shown in Figure 3.19), *land use* is shown in Figure 3.20, and *ownership patterns* are shown in Figure 3.21. *Land tenure* indicates whether the owner is a full-time, part-time, or tenant farmer. *Land use* is the percentage of the farm that is in rangeland and pastureland. *Ownership pattern* is an average of standardized

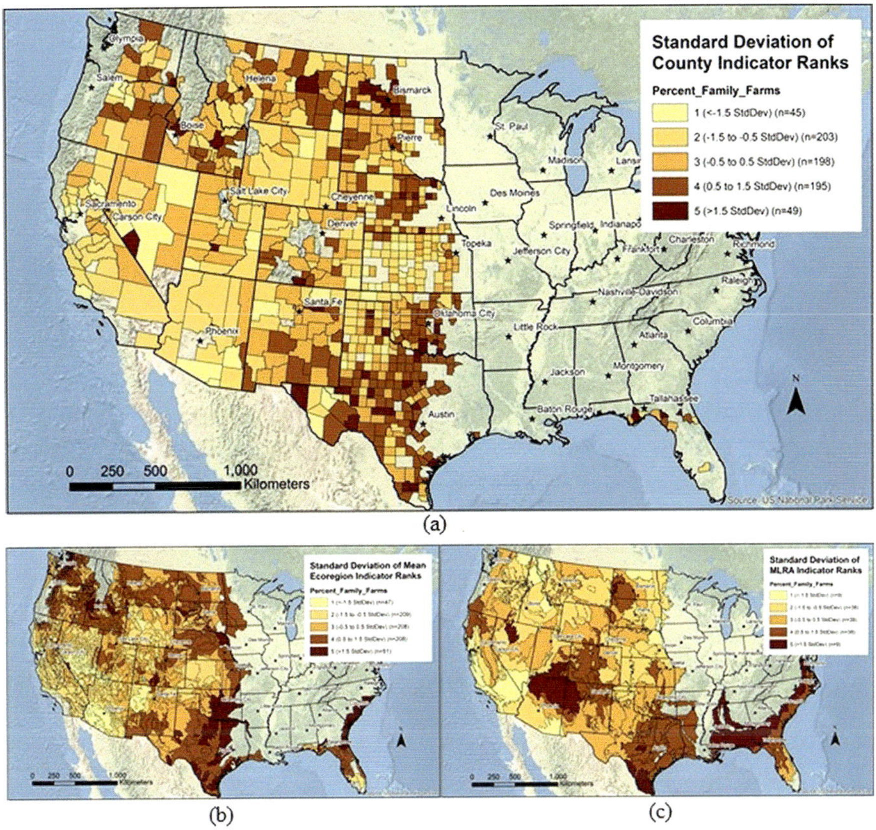

FIGURE 3.16
Percent of family farms of all Farms by (a) County, (b) Ecoregion, and (c) MLRA.

scores for occupation, gender, place of residence, off-farm work, age, internet availability, and the number of people in households.

Population Pyramid and Population Change

The *population pyramid and population change* indicator was divided into three separate population indicators. Each was calculated using the American Community Survey five-year estimates (US Census Bureau, 2020b). Population change was calculated as the *percentage change in estimated population from 2013 through 2017* (Figure 3.22). Given that population change in rangeland counties varies widely, we assume that population growth is more desirable and positively related to sustainability than is population decline.

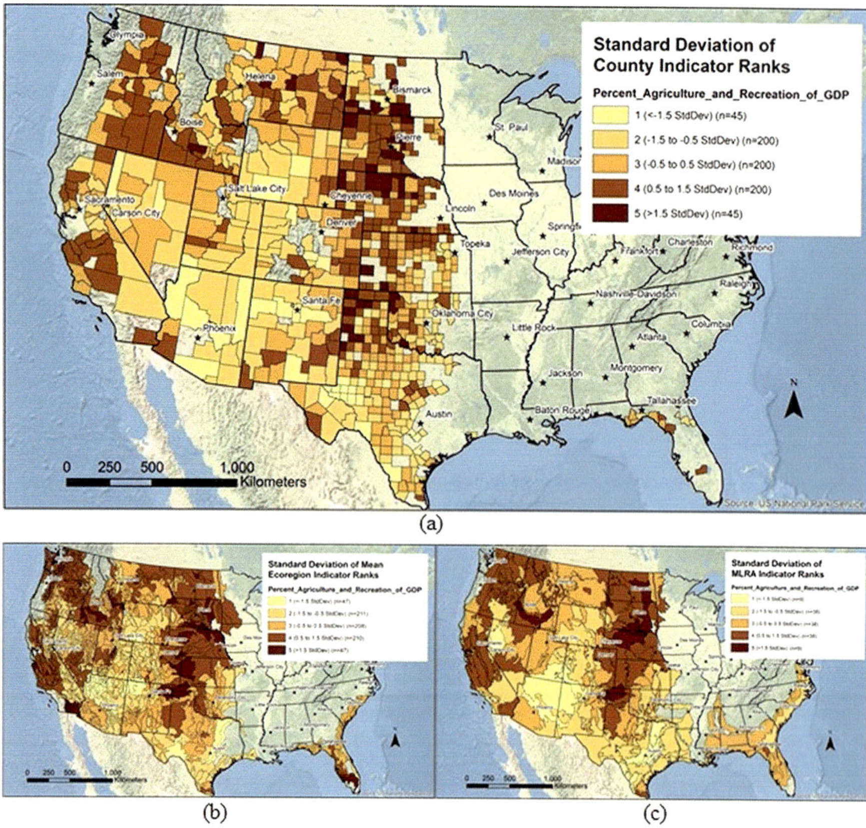

FIGURE 3.17

Value produced by agriculture and recreation industries as percent of total economic output by (a) County, (b) Ecoregion, and (c) MLRA.

Two additional indicators were created to address the original concept of the population pyramid. First, the *percent of population over the age of 65* was calculated (Figure 3.23). This is a measure of the proportion of the population that is aging, indicating the necessity for both social structure and economic resources to address the needs of this group. For example, a high percentage of people over the age of 65 with sufficient retirement earnings presents a different social scenario than the same percentage relying on Social Security for survival. Second, the *percent of population aged 18 years and under* was calculated (Figure 3.24). A high percentage of children presents different challenges to communities than does an aging population. As with other

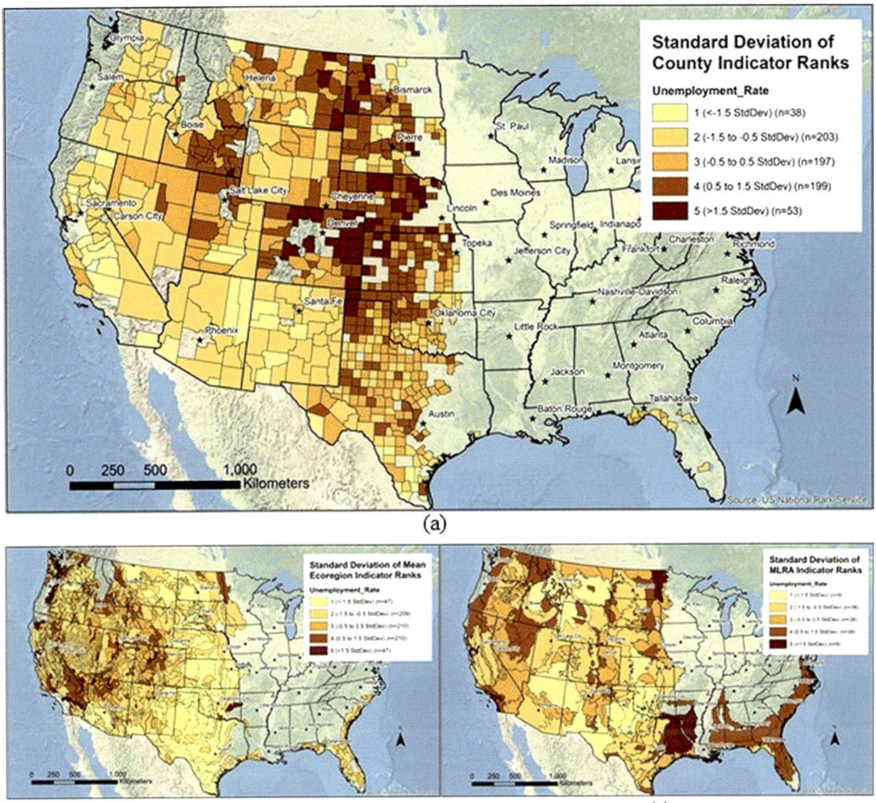

FIGURE 3.18
Unemployment rate by (a) County, (b) Ecoregion, and (c) MLRA.

indicators, these were standardized into a range of 0 to 100 and ranked for each county in the analysis.

Population growth is not distributed evenly across rangeland counties (Figure 3.22). However, it does follow known patterns. Counties with percent population change above the mean are generally in two groups. First, one group is experiencing natural growth and in-migration for social and economic reasons. This can be seen in growth associated with urban areas such as Phoenix, Denver, and Salt Lake City. Second, other counties grew because of specific economic shifts. Western North Dakota and the Permian Basin region of Texas indicate population growth associated with the expansion of domestic oil production during the time period used.

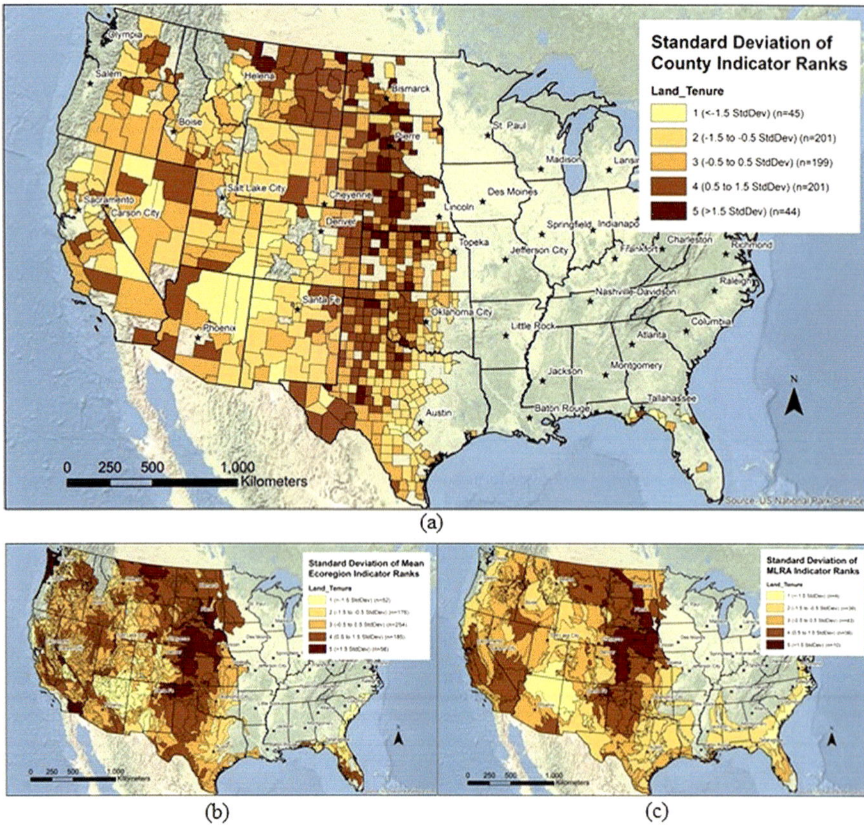

FIGURE 3.19
Land tenure by (a) County, (b) Ecoregion, and (c) MLRA.

Two patterns can be discerned from Figure 3.23. Both concern those rangeland counties whose percent population over 65 years was more than one standard deviation above the mean. First, many counties in this category are in regions known as retirement destinations, such as Phoenix, Las Vegas, and central Oregon. Second, other counties are in areas more likely to experience aging-in-place due to out-migration and a lack of natural replacement, such as the upper Great Plains. At the other end of the distribution, some rangeland counties have a percent population over 65 years more than one standard deviation below the mean. Most notable in this group are the counties in the Wasatch Front of Utah. This area is experiencing some of the highest population growth in the western United States.

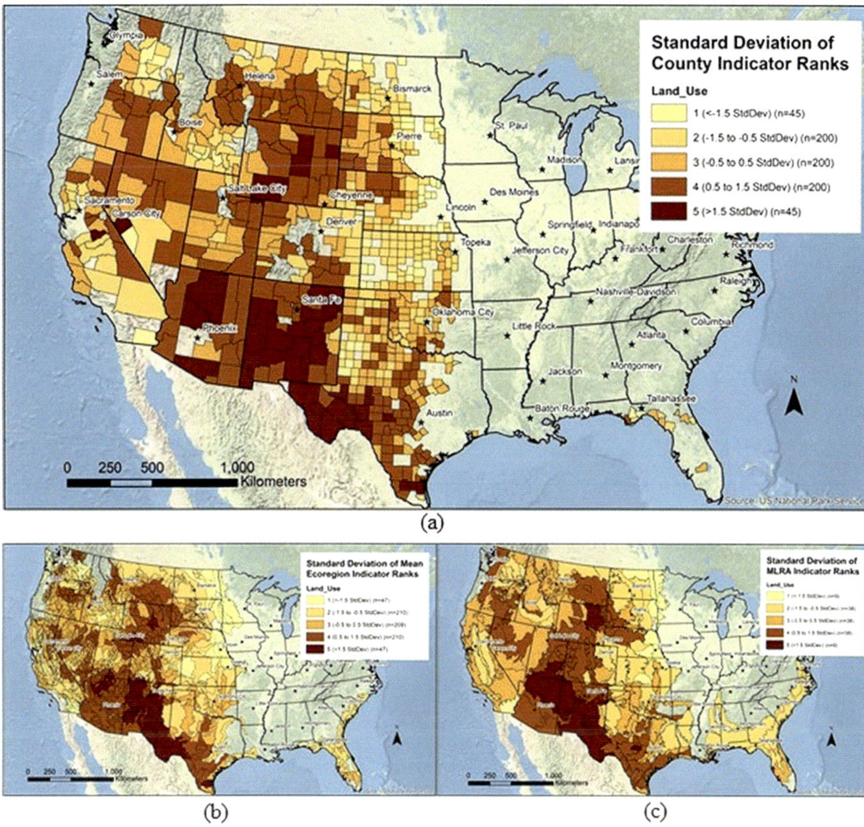

FIGURE 3.20
Land use by (a) County, (b) Ecoregion, and (c) MLRA.

Figure 3.23 is the converse of Figure 3.24. Rangeland counties with a percent of their population under the age of 18 that was above the mean mirror some of those counties with lower rankings for population over 65 years of age. Good examples are the Wasatch Front of Utah and the counties adjacent to the Denver area.

Income by Work Location versus Residence

Income by work location versus residence measures the extent to which residents of a county have earnings from employment within the county as opposed to working in other counties and returning to reside in the county in question

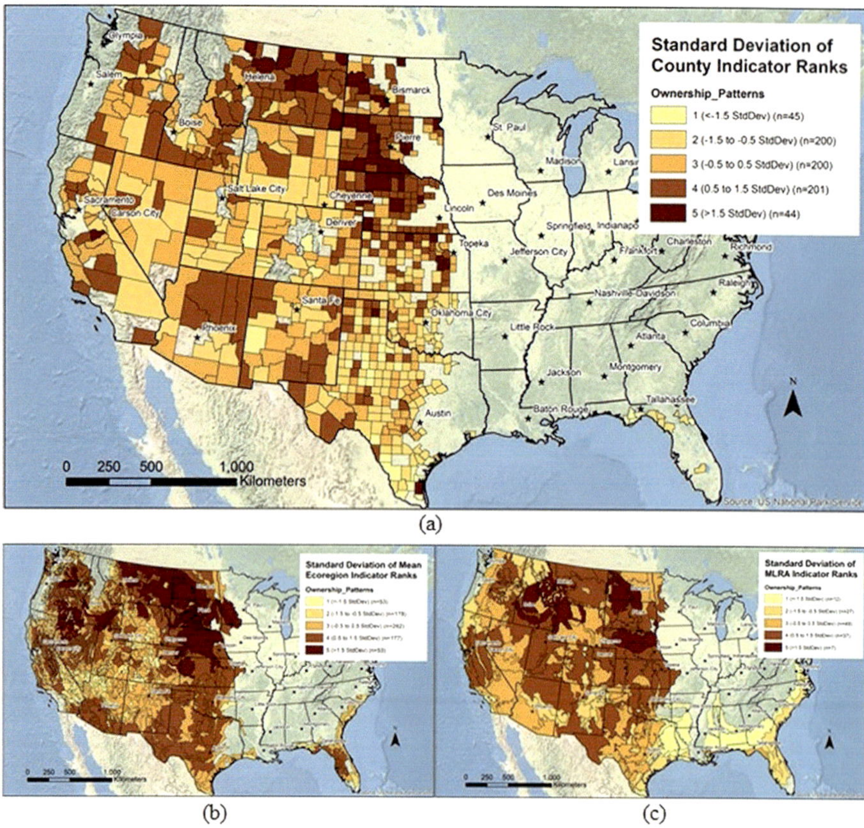

FIGURE 3.21
Ownership patterns by (a) County, (b) Ecoregion, and (c) MLRA.

(Figure 3.25). The earnings data for each county is adjusted for this commuting situation, resulting in either a positive or negative estimate of earnings adjustment. A positive number indicates that earnings are overwhelmingly generated within the county. A negative number indicates that most of these earnings are generated outside of the county.

Income by work location versus residence indicates the degree to which people live in one county and earn their living somewhere else. This can measure two, often simultaneous, forces. First, adjacency to another county allows for commuting patterns that raise the adjusted earnings in the home counties of commuters. Second, growth and income disparities raise the cost of living in

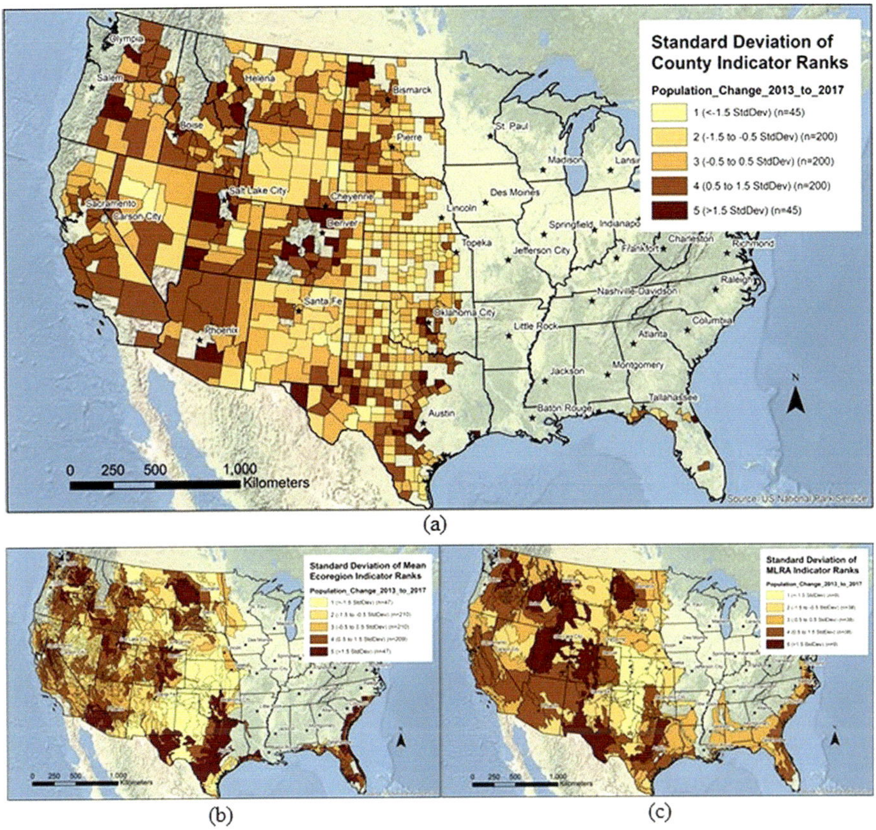

FIGURE 3.22
Percentage change in estimated population from 2013 through 2017 by (a) County, (b) Ecoregion, and (c) MLRA.

a county to the point where much of the workforce cannot afford to live there and is pushed into adjacent counties. In either case, the impacts are important. For example, the availability of off-farm employment via commuting can be significant. Also, other economic spillover effects are likely when commuters spend their earnings in their home county. Southern California, Phoenix and Tucson in Arizona, Denver, and even southwest Idaho display patterns of commuting and earning flows above the mean when compared to other rangeland counties. Counties below the mean can have high transfer payments and unearned income, such as pension and other retirement-related payments.

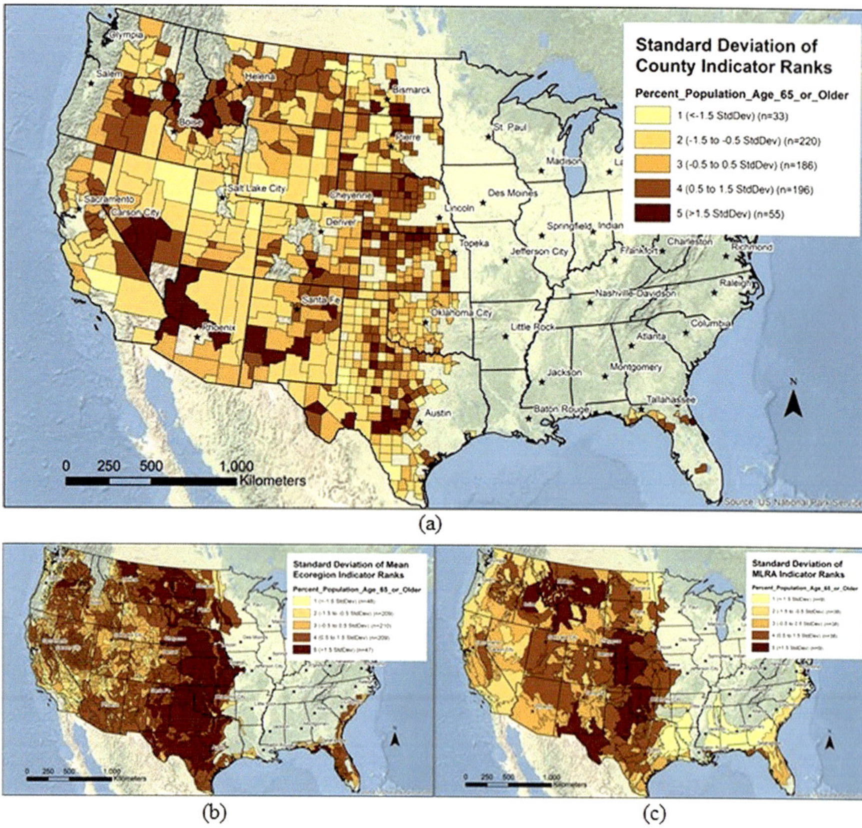

FIGURE 3.23
Percent of population over the age of 65 by (a) County, (b) Ecoregion, and (c) MLRA.

What Do the Social and Economic Indicators Indicate?

With the above as a prelude, we can take a quick look at how the indicators stack up and see what they tell the reader. We will start with an overall look at the social and economic indicators and then move to see what they mean in terms of the Sustainability Triangle in the following chapter.

What Do the Social Indicators Mean by Themselves?

The ranks for each county on each indicator are averaged to create a state cumulative average for that indicator. These rank scores for indicators by state are summed and presented in Figure 3.26. Each color shows the average

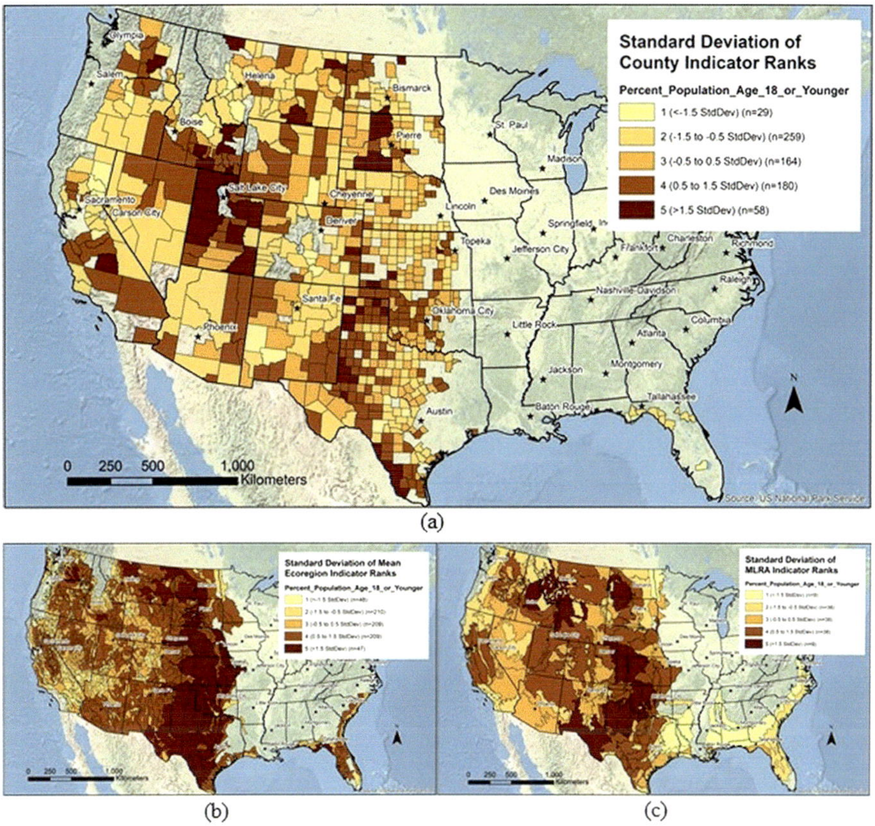

FIGURE 3.24
Percent of population aged 18 years and under by (a) County, (b) Ecoregion, and (c) MLRA.

rank for that indicator for each state. The total height of the bar indicates the total, overall cumulative rank on social indicators for a state. If a state were to be ranked at 100 for every indicator, the maximum possible score would be 800. Higher scores indicate progress toward social sustainability. The final distribution of these scores ranges from a low of 301 for New Mexico to a high of 450 for Idaho.

Social indicator ranks vary within states. As an example, Figures 3.27 and 3.28 rank scores for each social indicator in each rangeland county in Idaho and New Mexico, respectively. Idaho's rank sums range from a low of 274 to a high of 571 out of a possible 800. New Mexico's rank sums range from a low of 217 to a high of 453 out of a possible 800. Idaho ranks higher than New Mexico on our cumulative measure of social sustainability. To put it another way, Idaho has a higher cumulative rank on our social indicator measures

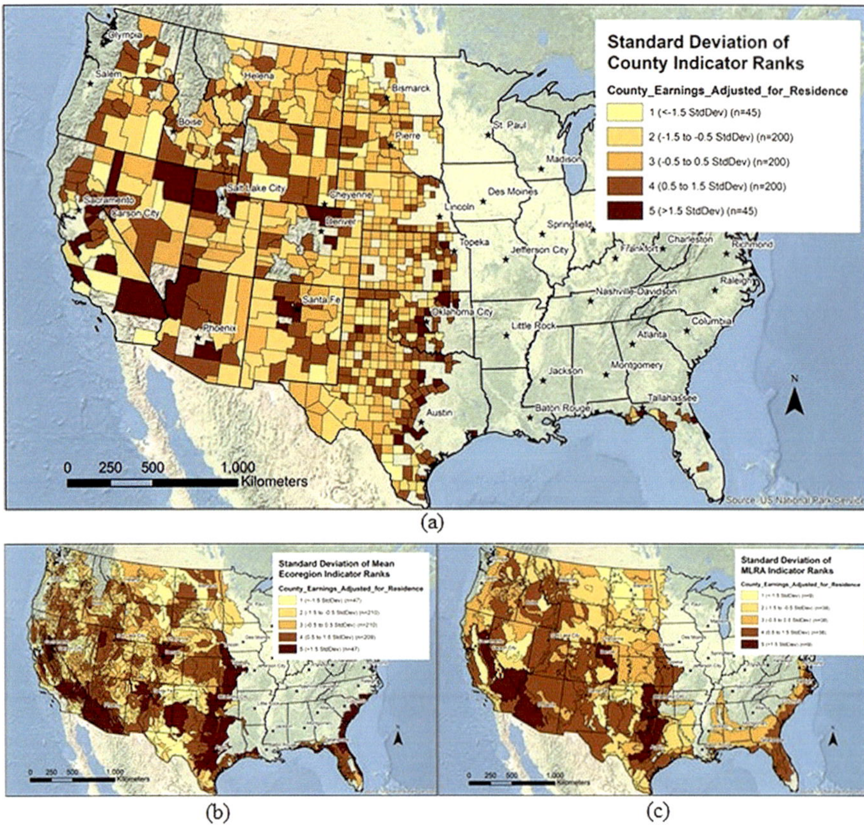

FIGURE 3.25
Income by work location versus residence by (a) County, (b) Ecoregion, and (c) MLRA.

relative to other states based on a comparison of all 690 rangeland counties. Figure 3.29 shows the standard deviation categories for all rangeland counties for standardized ranks of social indicator cumulative scores. We can easily identify pockets of social well-being relative to an overall mean, as well as pockets of possible social concern. We can examine county ranks on social indicators relative to all other rangeland counties and see which rank higher or lower than others. Whether any county is socially sustainable is a conclusion with which readers can wrestle.

What Do the Economic Indicators Mean by Themselves?

States and counties vary in how they rank against each other by economic indicators. Figure 3.30 shows how each state compares when the average of

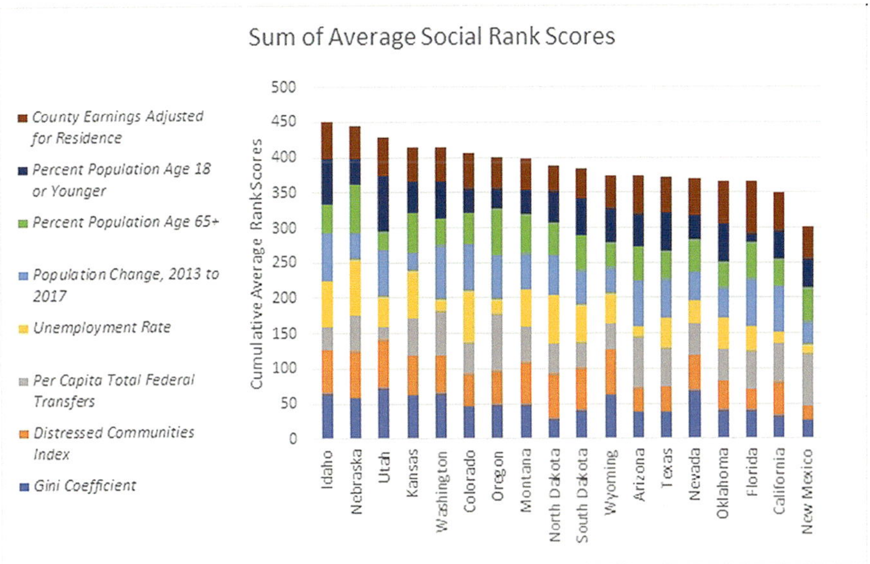

FIGURE 3.26
Cumulative Social Indicator Standardized Rank Scores by State Ranging from 301 to 450 out of 800.

each county economic indicator standard score is added together. As shown, South Dakota (1,024) scores the highest, and Florida (549) scores the lowest. Note that the maximum possible sum would have been 1,700 if a state scored the highest on every indicator. Compared to the other states, South Dakota ranks high (top quartile) in 10 of the 17 indicators and low (bottom quartile) in 2 of the 17. Florida ranks high in 2 of 17 and low in 10 of 17.

Even within those states, there are differences among the counties. Figure 3.31 shows the South Dakota rangeland counties with summed indicator scores ranging from 705 to 1,235. Figure 3.32 shows Florida rangeland counties with summed indicator scores ranging from 397 to 774. In fact, only one South Dakota county scores lower than the highest Florida county.

If we look at the percentage of counties in each state by whether they fall in the highest or lowest quartile, you get a figure such as Figure 3.32. In this figure, the counties in each state determined whether they fall in the lowest or highest quartile and then the average indicator scores for just those counties are calculated. South Dakota has the lowest cumulative percentage of counties in the lowest quartile and one of the highest cumulative percentages of counties in the highest quartile. Similarly, Florida has the highest cumulative percentage of counties in the lowest quartile and the lowest percentage in the highest quartile.

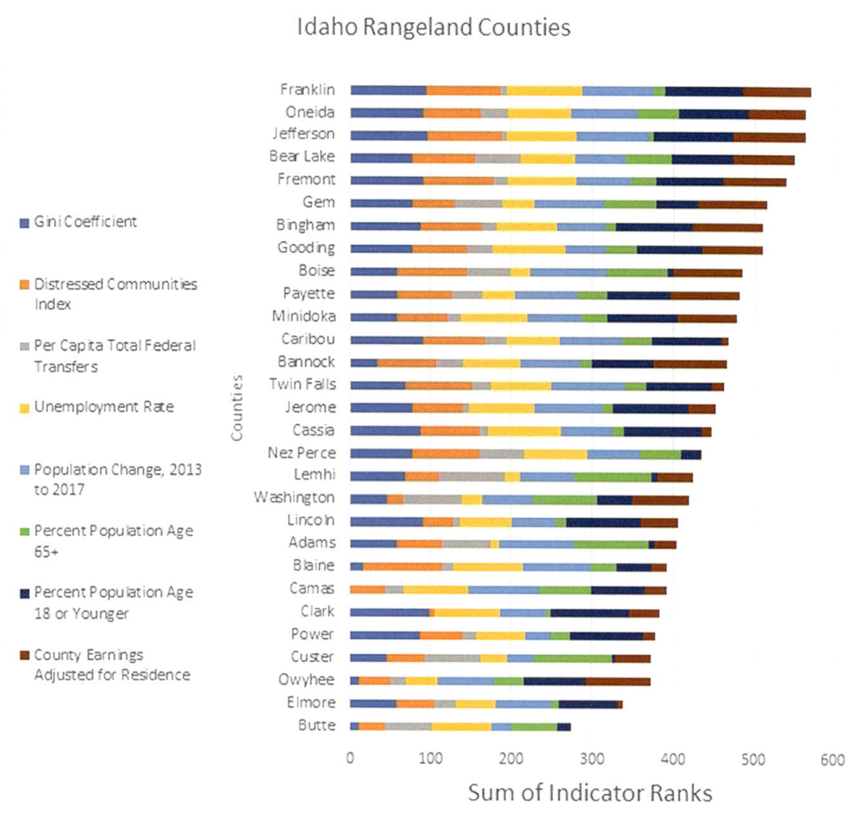

FIGURE 3.27
Idaho County social indicator standardized rank sums ranging from 274 to 571 out of 800.

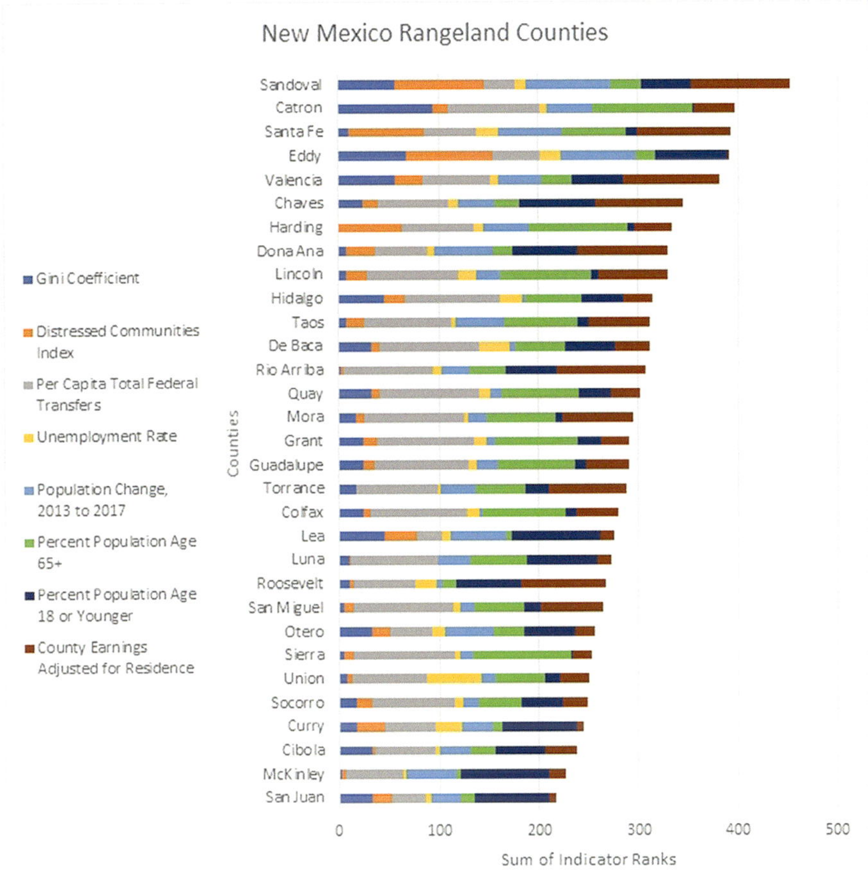

FIGURE 3.28
New Mexico County social indicator standardized rank sums ranging from 217 to 453 out of 800.

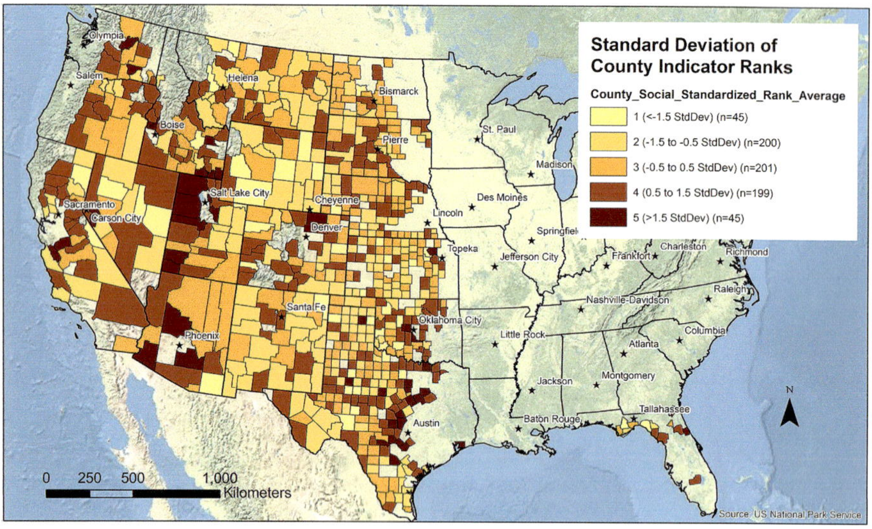

FIGURE 3.29
Distribution of standardized rank average scores for social indicators by standard deviation categories.

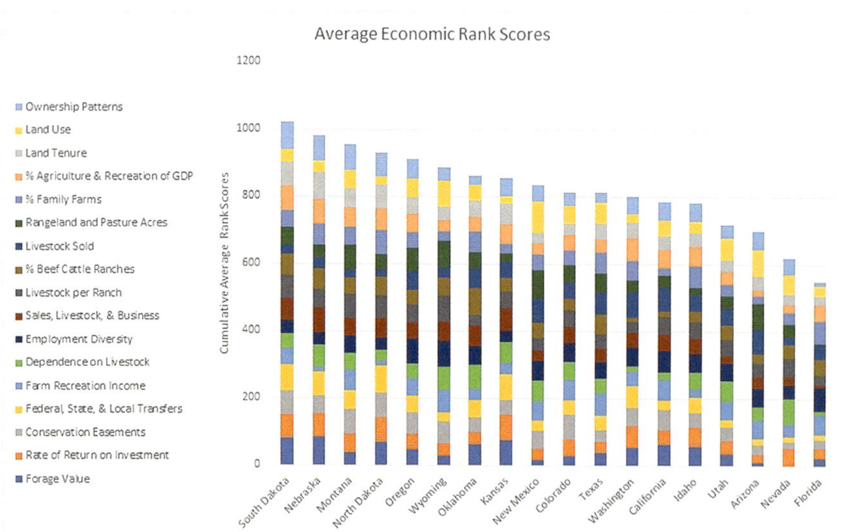

FIGURE 3.30
Sum of average economic indicator standardized rank scores by state ranging from 549 to 1,024 out of 1,700.

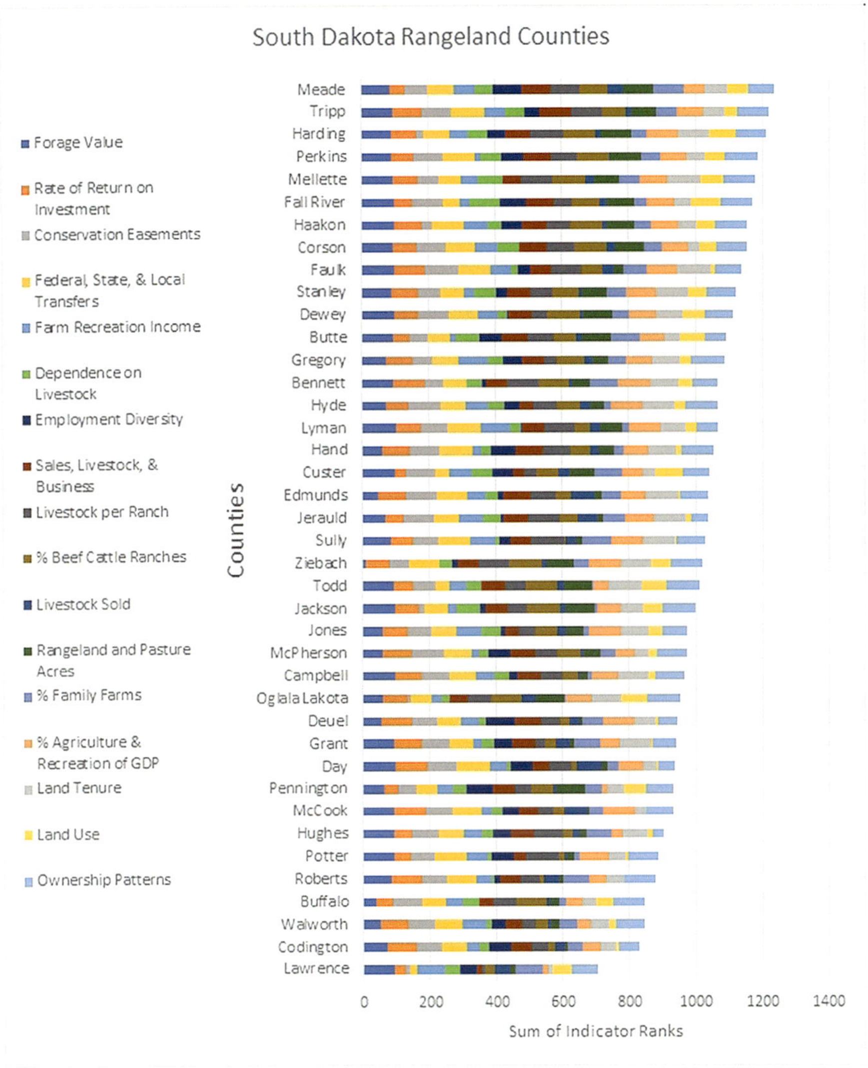

FIGURE 3.31
South Dakota County economic indicator standardized rank sums ranging from 649 to 1,035 out of 1,700.

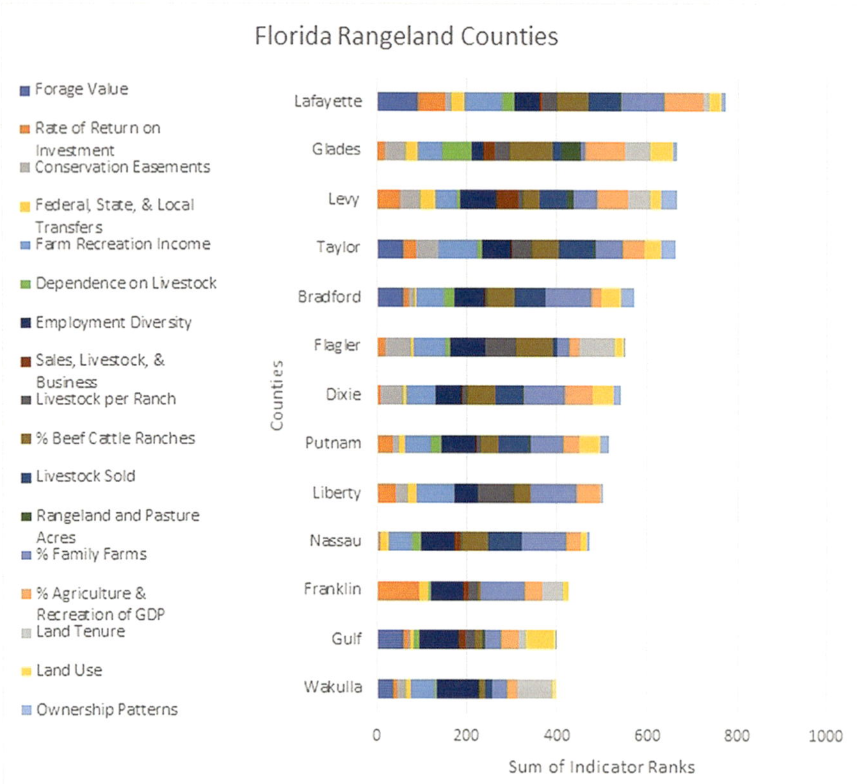

FIGURE 3.32
Florida Rangeland County economic indicator standardized rank sums ranging from 397 to 774 out of 1,700.

Does this mean that South Dakota would be considered more economically sustainable compared to Florida? Perhaps. The data show that South Dakota ranks higher compared to all other rangeland nonurban counties. Whether it is better to rank in the highest quartile in such things as forage value, rate of return on investment, conservation easements, federal agriculture transfers, value produced by agriculture and recreation, and ownership patterns is up for discussion.

We can see how the rangeland counties are distributed based upon their rank categories (Figure 3.33). It appears that counties in the Northern Great Plains and the Northern Great Basin regions are doing generally better than the rest of the rangeland counties from an economic perspective, as indicated by the darker colors on the map. Counties in the Southern Great Basin, desert southwest, and Florida appear to be doing less well economically compared to the other rangeland counties.

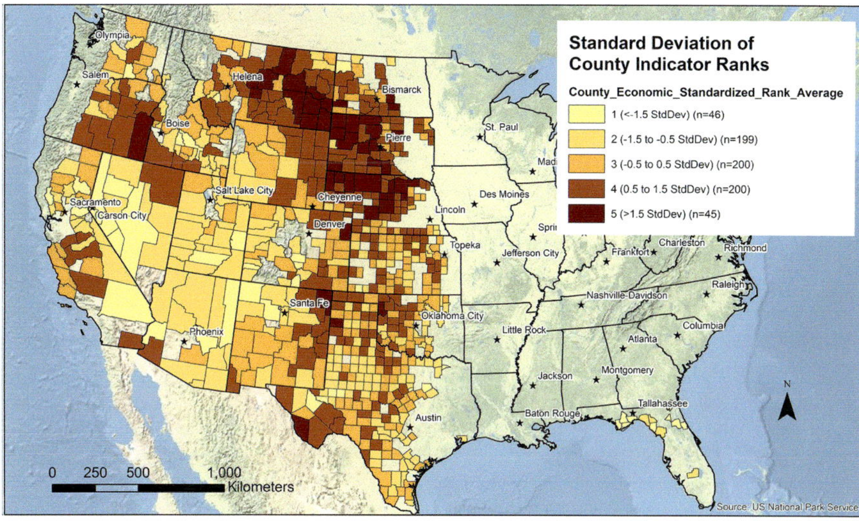

FIGURE 3.33
Distribution of standardized rank average scores for economic indicators by standard deviation categories.

Literature Cited

[BEA] Bureau of Economic Analysis, 2020. Regional GDP & personal income. Gross domestic product, CAGDP2: GDP in current dollars by county and MSA. https://apps.bea.gov/regional/downloadzip.cfm (accessed 3 August 2020)

[BLS] Bureau of Labor Statistics, 2020. Quarterly census of employment and wages; county high-level. https://www.bls.gov/cew/downloadable-data-files.htm (accessed 28 January 2020)

Economic Innovation Group, 2018. *From great recession to great reshuffling: charting a decade of change across American communities.* Washington, DC. https://eig.org/wp-content/uploads/2018/10/2018-DCI.pdf (accessed 3 September 2020).

ESRI, 2019. *ArcGIS ArcMap: Release 10.7.1.* Environmental Systems Research Institute, Redlands, CA.

Fox, W.E., McCollum, D.W., Mitchell, J.E., Swanson, L.E., Kreuter, U.P., Tanaka, J.A., Evans, G.R., Heintz, H.T., 2009. An integrated social, economic, and ecologic conceptual (ISEEC) framework for considering rangeland sustainability. *Society and Natural Resources* 22, 593–606.

Isserman, A., 2005. In the national interest: defining rural and urban correctly in research and public policy. *International Regional Science Review* 28, 465–499. https://doi:10.1177/0160017605279000.

Janker, J., Mann, S., 2020. Understanding the social dimension of sustainability in agriculture: a critical review of sustainability assessment tools. *Environmental Development and Sustainability* 22, 1671–1691. https://doi.org/10.1007/s10668-018-0282-0.

Land, K., Michalos, A., Sirgy, M., 2012. *Handbook of social indicators and quality of life research*. Springer, Netherlands.

Lichter, D., Ziliak, J., 2017. The rural-urban interface: new patterns of spatial interdependence and inequality in America. *ANNALS, American Academy of Political and Social Science* 672, 6–25.

Mitchell, J.E. (Ed.), 2010. *Criteria and indicators of sustainable rangeland management*. University of Wyoming Extension Publication Number SM-56, Laramie, WY, p. 227.

[NASS] National Agricultural Statistics Service, 2019. 2017 census of agriculture. https://www.nass.usda.gov/Publications/AgCensus/2017/index.php (accessed 20 April 2020).

[NASS] National Agricultural Statistics Service, 2020a. Quick stat tools. https://www.nass.usda.gov/Quick_Stats/index.php (accessed 20 April 2020).

[NASS] National Agricultural Statistics Service, 2020b. Rent, cash, pastureland – expense, measured in $/acre – value. https://www.nass.usda.gov/Surveys/Guide_to_NASS_Surveys/Cash_Rents_Cy_County/ (accessed 11 August 2020).

Slack, T., Jensen, L., 2020. The changing demography of rural and small-town America. Population Research and Policy Review, Published online: 14 September 2020. https://doi.org/10.1007/s11113-020-09608-5

[SRR] Sustainable Rangelands Roundtable, 2003. Criteria and indicators for sustainable rangelands. A first approximation report by the Sustainable Rangelands Roundtable. https://www.sustainablerangelands.org/projects/criteria-and-indicators/first-approximation-report/ (accessed 27 October 2020).

Tanaka, J., Torell, L.A., Swanson, L.V., Brunson, M., McCollum, D., Heintz Jr., H.T., 2009. Chapter V: Indicators for maintenance and enhancement of multiple economic and social benefits to current and future generations, in: *Sustainable rangeland roundtable, first approximation report*. University of Wyoming. http://www.sustainablerangelands.org/reports/2003Report/Chapter05.pdf (accessed 3 September 2020).

Tickamyer, A., Sherman, J., Warlick, J., 2017. *Rural poverty in the United States*. Columbia University Press, New York.

US Census Bureau, 2020a. American community survey, 2017. Data table B19083, Gini index of income inequality, 2017 American community survey 1-year estimates. https://data.census.gov/cedsci/table?q=B19083&tid=ACSDT1Y2017.B19083&hidePreview=false (accessed 3 September 2020).

US Census Bureau, 2020b. American community survey 5-year estimates – public use microdata sample (2017). https://data.census.gov/mdat/#/ (accessed 3 September 2020).

Wunderlich, G., 2016. *Rationalizing rural area classifications for the economic research service: a workshop summary*. The National Academies Press, Washington, DC. https://doi.org/10.17226/21843

4

Assessing Rangeland Sustainability

Rangeland sustainability specifically and sustainability in general are described in concept, but it is difficult to know if they are achieved. As the SRR described at their inception, we are not here to tell the world that we are sustainable but rather to identify the data that can lead to informed discussion. Personal preferences and values frame how an individual, group, or organization looks and interprets the same data. This chapter is an attempt to stimulate discussion as people look at that data as a way to organize the social, economic, and ecologic information.

RANK SCORING

For all social, economic, and ecological indicators, raw data for each indicator were ranked among the 690 rangeland counties. Counties are categorized by how many standard deviations they are from the mean of all counties. All analysis and discussion in this chapter are based on counties, ecoregions, MLRAs, or congressional districts (CD) compared to all of the other counties, ecoregions, MLRAs, or CD, respectively, using the standard deviation categories.

Displaying Rangeland Sustainability

The tripartite way of organizing sustainability information as social, economic, and ecological (environment) is often depicted as a Venn diagram illustration (Figure 4.1). Conceptually, where the three circles overlap is considered to be the sustainable part of the diagram. While an effective way to visualize the issue, it does not necessarily indicate where you sit in the diagram.

While the Venn diagram approach gives a conceptual view, with each part being equal and of equal weight in the discussion, it does not provide guidance on how much overlap there is or where individual and societal values lie.

Another way to look at sustainability would be balancing plates on a fulcrum (Figure 4.2). In this depiction, the idea would be that each set of indicators is a plate, and those plates sit on a tray balanced on a fulcrum. As society

DOI: 10.1201/9781003279747-4

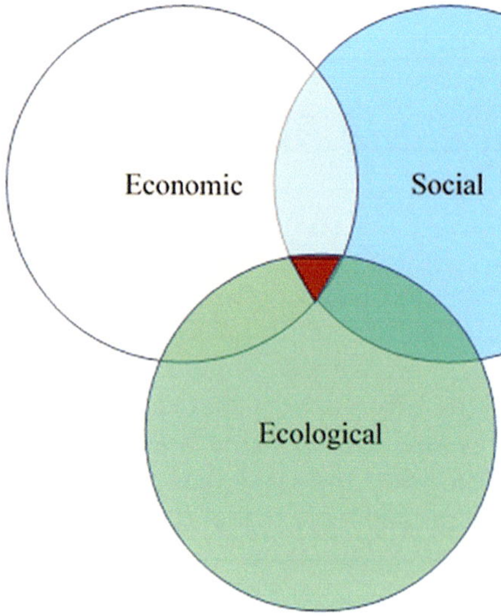

FIGURE 4.1
Traditional Venn diagram representation of sustainability. The red triangle represents that part of the world where economic, social, and ecological aspects are in balance and is viewed as being sustainable.

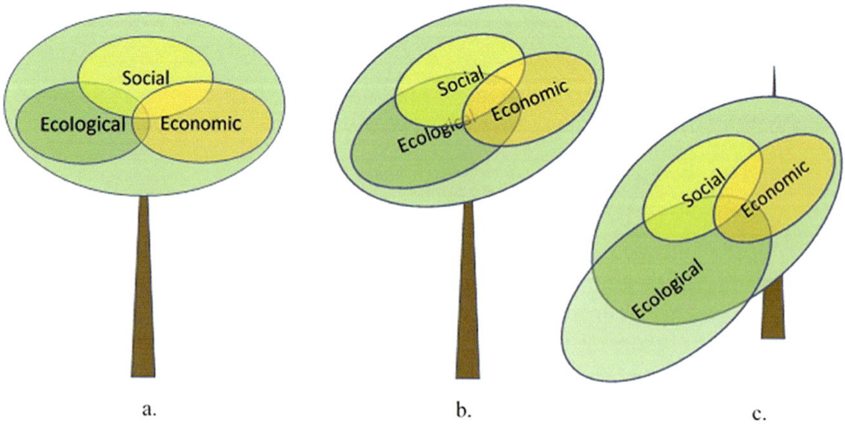

FIGURE 4.2
A balanced plate depiction of rangeland sustainability. (a) The system is in balance, (b) more weight (individual or societal value) is put on the ecological system and the tray sways in that direction but is still in balance (sustainable), and (c) too much weight is put on the ecological system and the tray falls off the fulcrum (unsustainable).

or individual values change as to what is more important, the tray would tilt in that direction, but the plates stay on. In this case, as long as the tray stays on the fulcrum, the rangeland system would be considered sustainable (roughly in balance). If values put too much weight on one of the plates, the tray falls off the fulcrum and is no longer sustainable.

While this depicts sustainability more graphically than the Venn diagram in terms of what is sustainable or not sustainable, it still does not lead to a way to determine how large each plate is or whether the tray is still on the fulcrum. In other words, there is no way to quantify whether a system is sustainable or not.

A third way to depict sustainability along the three dimensions is to develop and use something like the Soil Triangle (Figure 4.3). We propose to put social, economic, and ecological scores along each side of a triangle and develop criteria for determining whether it can be said that a rangeland system is sustainable or unsustainable. It may also depict whether one or more of the indicator sets is given more or less weight than the others. While the Soil Triangle is based on the percentage of sand, silt, and clay in a soil sample and always has to add to 100%, there is no analogue for social, economic, and ecological systems.

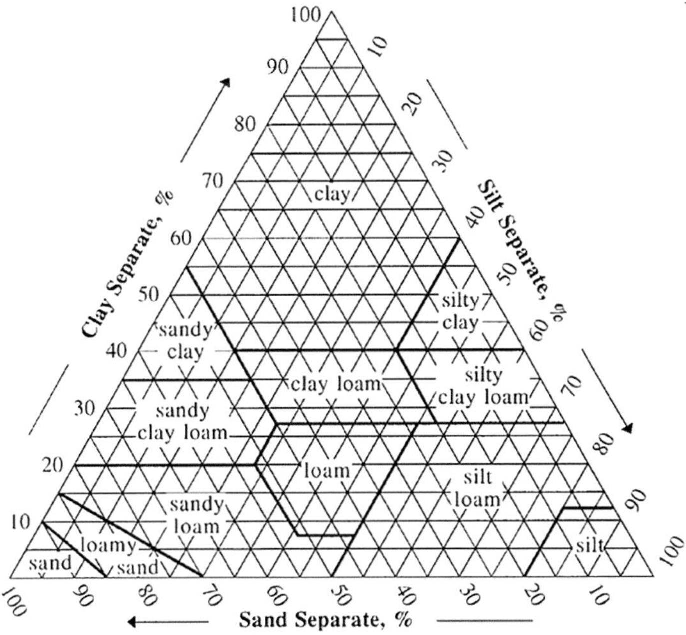

FIGURE 4.3
Soil Triangle (NRCS, 2020).

So, the challenge is to develop a similar set of data for the ecological, social, and economic indicators. One potential method is to use what is called the Human Footprint Model (Sanderson et al., 2002; Leu et al., 2008; Woolmer et al., 2008; Deutsch, 2011). In essence, this model seeks to score each criterion based on the impact it might have on the ecosystem. These scores are then normalized and added together to get an overall score that can be put on a 0 to 100 basis (Figure 4.4). The scores, in this model, can then be displayed on

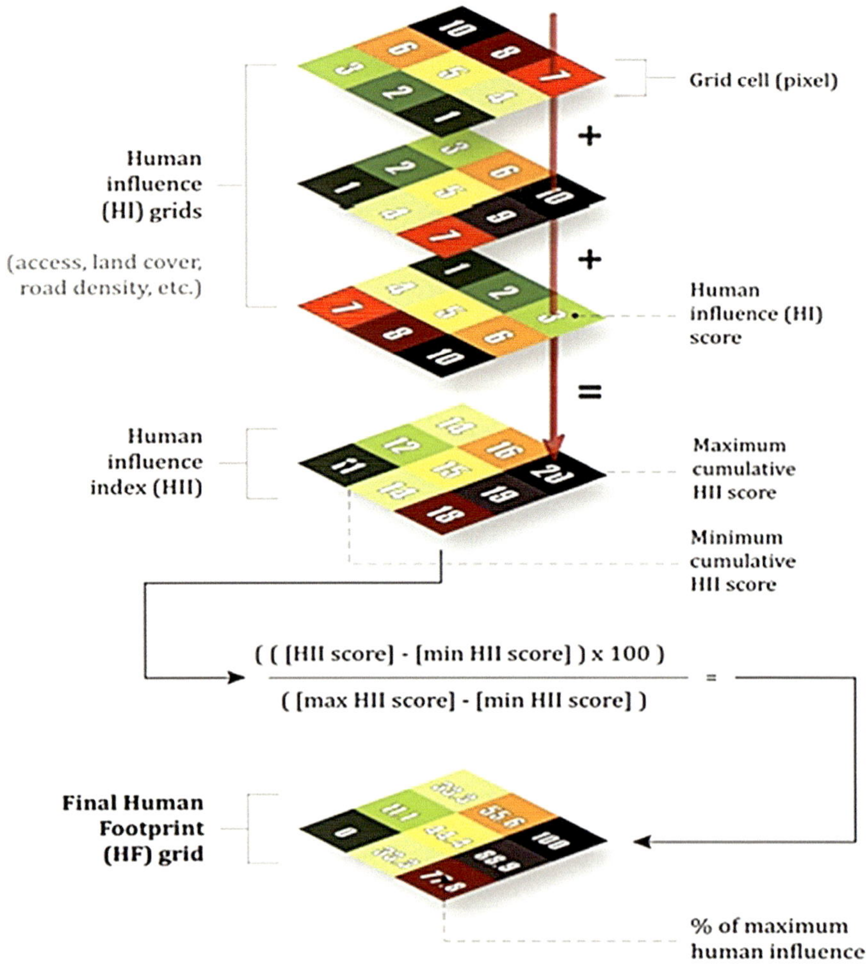

FIGURE 4.4
Human footprint process (Deutsch, 2011).

a GIS map to show how human impacts change over time, or with the introduction of a stressor to the system (e.g., a new road going through a natural area). Conceptually, this is what we are attempting to show with rangeland sustainability.

The indicators developed by the SRR process are a combination of quantitative and qualitative measures. The quantitative indicators lend themselves well to being transformed into a standard score. Conceptually, a Sustainability Triangle (Figure 4.5) would be set up similar to the Soil Triangle. What is not clear is where the demarcation lines would be drawn, nor what the sustainable combination of indicator scores would be. While Figure 4.4 shows conceptually what such a Sustainability Triangle might look like, more research needs to be done to place the lines and descriptions.

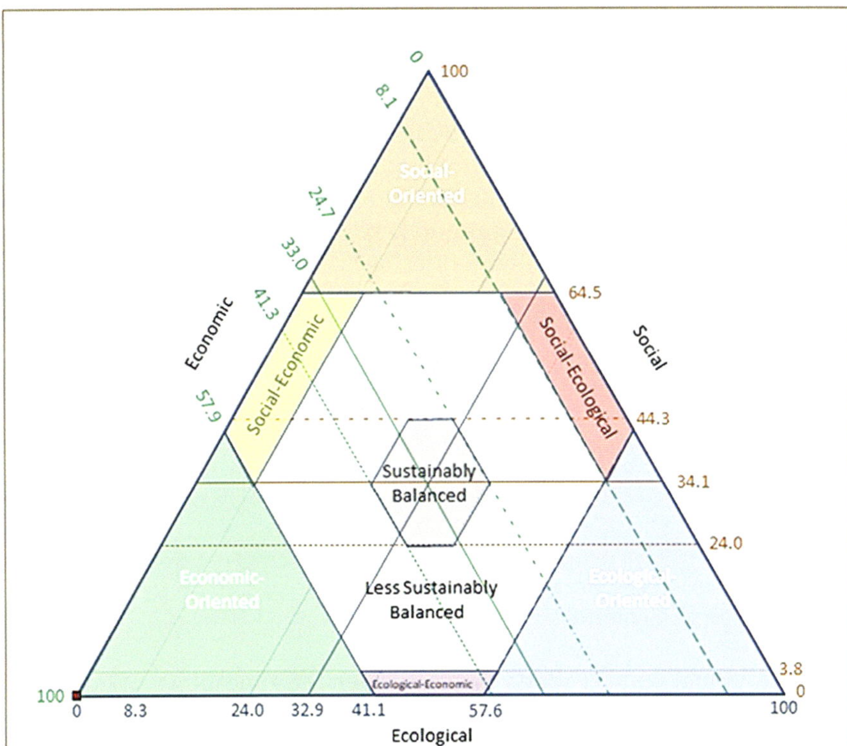

FIGURE 4.5
Rangeland Sustainability Triangle. Dividing lines based on plus or minus 0.5 or 1.5 standard deviations from the mean. Mean ranks are shown with solid lines and the standard deviation lines in dots and dashes.

The implication in Figure 4.5 is that social, economic, and ecological indicators could all have a score of 33 1/3 and be in the sustainably balanced range (the central hexagon). We selected from −0.5 to 0.5 standard deviations for each category to define the sustainably balanced range. Similarly, we selected −1.5 to 1.5 standard deviations to define the less sustainably balanced range. In the Human Footprint Model, the assumption is made that every indicator is given equal weight in coming up with a total score. It is relatively easy to apply different weights to different indicators to show how values may change the perception of sustainability.

Sustainability Triangle

Each indicator of rangeland sustainability is ranked. For counties, these would be from 1 to 690 with ties having the same rank. A rank of 1 is considered to be the lowest (worst) scored county and 690 the highest (best). For each social, ecological, or economic category, the ranks for all indicators are averaged to get an average total rank score for each county. Since each criterion does not have the same number of indicators in total and across time, the average accounts for that variation. With those average scores for ecological, economic, and social categories, we can then plot it on a Rangeland Sustainability Triangle. We have chosen to use a ternary plot algorithm developed by Advocate (2019).

It is necessary to develop a metric that ensures the sum of the three indicator sets is 100. This requires using a formula to re-score the indicator sets one more time such that

$$\text{Sustainability Score} = \sum S_i = 100,$$

where
S_i = Standardized score for category i, and
i = Ecological, Economic, Social.

Each individual score is added together. Then the allocation works as

$$S_i * 100/\left(\sum S_i\right).$$

And the point in the triangle can then be plotted.

The Rangeland Sustainability Triangle is set up so that the sustainability scores for all 690 counties have an average and a standard deviation

TABLE 4.1

Average Standardized Ranking and Standard Deviations for 690 Rangeland Counties for Social, Economic, and Ecological Criteria

	Social	Economic	Ecological
Average	34.1	33.0	32.9
Standard deviation	20.2	16.6	16.4

calculated (Table 4.1) for ecological, economic, and social indicators. On the triangle, a line is drawn for each criterion such that it represents the average sustainability score (solid lines in Figure 4.5). Additional lines are drawn for plus or minus 0.5 or 1.5 standard deviations. The hexagon in the middle (gray) represents all county scores that are within 0.5 standard deviations of the mean and labeled sustainably balanced (similar to Figure 4.2a). Likewise, the area within the larger hexagon (white) represents all counties that are between plus or minus 0.5 and 1.5 standard deviations from the mean and labeled less sustainably balanced (similar to Figure 4.2b). Each colored region on the triangle is given a descriptive name as indicated and represents different parts of the Venn diagram depiction (Figure 4.1).

Using the ternary plot algorithm, points within the Rangeland Sustainability Triangle for every county can be plotted as in Figure 4.6. The triangle is divided based on plus or minus 0.5 and 1.5 standard deviations from the mean for social, economic, and ecological standardized scores and are given names to represent where they fall (Table 4.2). As can be seen, the majority of the counties fall within the sustainably balanced or less sustainably balanced regions of the triangle. We interpret the sustainably balanced range to be those counties where ecological, economic, and social indicators are roughly in balance. The less sustainably balanced range are leaning in one direction a little more. For any of the other six regions of the triangle, we interpret those counties as out of balance for one reason or the other.

Additionally, it is possible to see how different groupings of counties would look using the MLRA (Figure 4.7), Küchler Ecoregions (KE; Figure 4.8), and CD (Figure 4.9) divisions. Figures 4.7–4.9 show how county values aggregated based on the percentage of a county in a division would look. There appears to be a smoothing effect that occurs, as the units are more compressed than for the individual counties. For each aggregation, raw data are ranked against the others in the division and standardized.

So how do we determine what the individual plotted points mean in terms of sustainability? What if the different criteria are given different weights? For example, what if a group determines that ecology should be worth twice

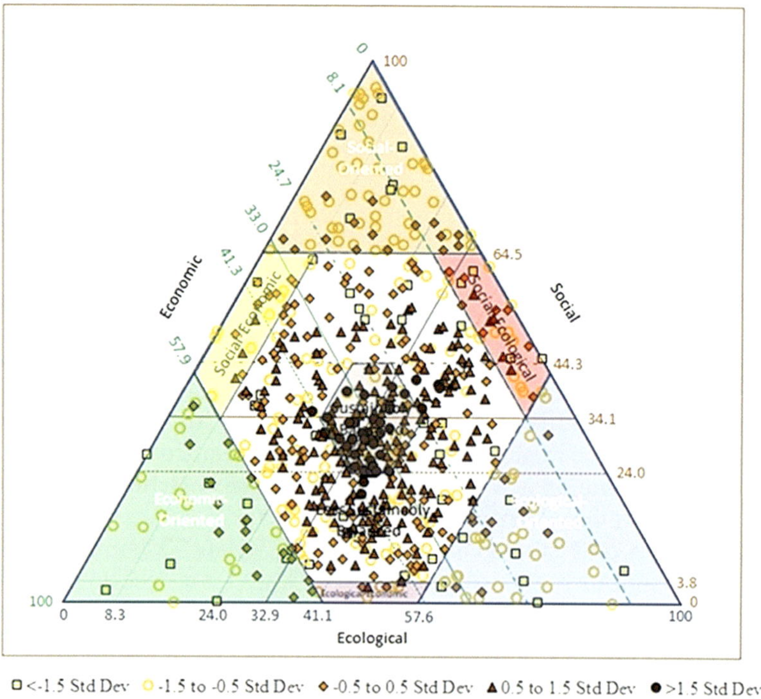

(a)

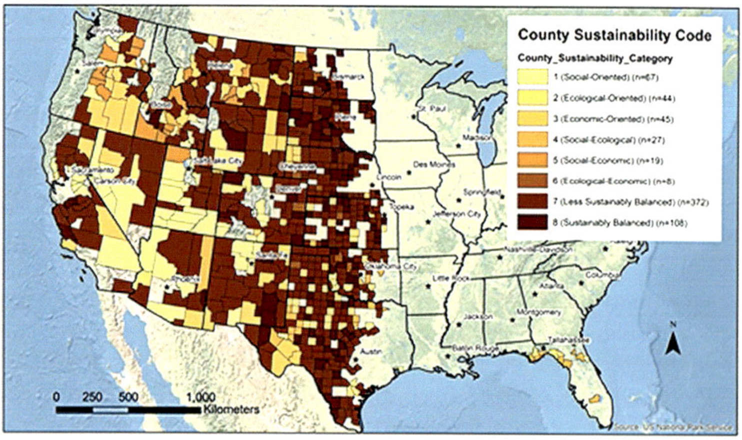

(b)

FIGURE 4.6
Sustainability Triangle (a) for All 690 Rangeland, Nonurban Counties Based on 2017 Data and the Map Distribution (b).

TABLE 4.2

Sustainability Triangle Region Categories, Names, and Ranges. Note That Every Point on the Sustainability Triangle Must Equal 100

Sustainability Category	Name	Range
1	Social-oriented	Social > 64.5
2	Ecological-oriented	Ecological > 57.6
3	Economic-oriented	Economic > 57.9
4	Social-ecological	Social < 64.5; Ecological < 57.6; Economic < 8.1
5	Social-economic	Social < 64.5; Ecological < 8.3; Economic < 57.9
6	Ecological-economic	Social < 3.8; Ecological < 57.6; Economic < 57.9
7	Less sustainably balanced	Social > 3.8 and < 24.0; Social > 44.3 and < 64.5; Ecological > 8.3 and < 24.0; Ecological > 41.1 and < 57.6; Economic > 8.1 and < 24.7; Economic > 41.3 and < 57.9
8	Sustainably balanced	Social > 24.0 and < 44.3; Ecological > 24.0 and < 41.1; Economic > 24.7 and < 41.3

what social and economics indicate? In that case, the process described previously can have weights added in such that

$$\text{Sustainability Score} = W_{\text{Ecological}}S_{\text{Ecological}} + W_{\text{Social}}S_{\text{Social}} + W_{\text{Economic}}S_{\text{Economic}} = 100,$$

where W_i = subjective or objective weight given to category i.

Let us see how all this would work with scores. Table 4.3 shows a hypothetical result. Figure 4.10 illustrates how the unweighted and weighted results would look on the Sustainability Triangle for a representative county. When the ecological score is given twice the weight of social or economic scores, the position on the Sustainability Triangle will change. If the boundary lines are on the appropriate scores, this worldview would move this example from within the less sustainably balanced section to the ecological-oriented category. While this would be better for those whose values lean toward ecology and the environment, it may not bode as well for the economy or rural communities. In this case, the results are sensitive to the weights, but other cases may differ.

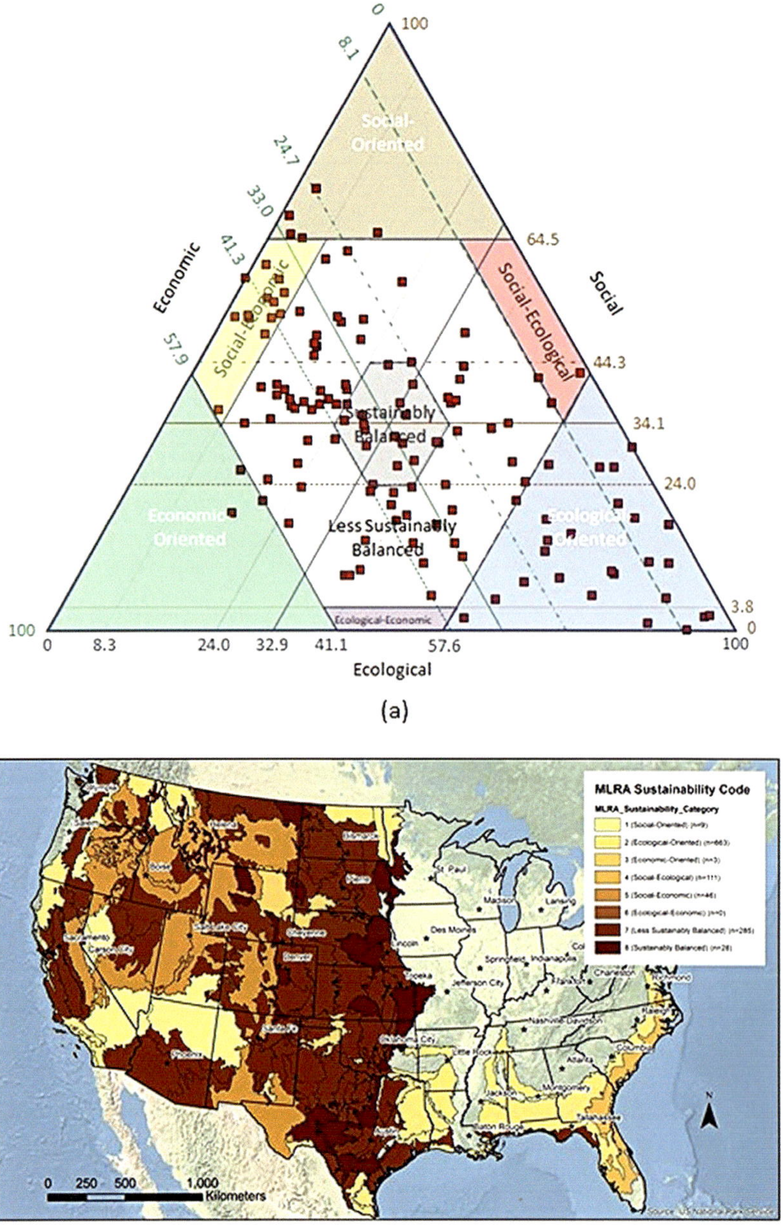

FIGURE 4.7
Sustainability Triangle (a) for MLRA Based on 2017 Data and the Map Distribution (b).

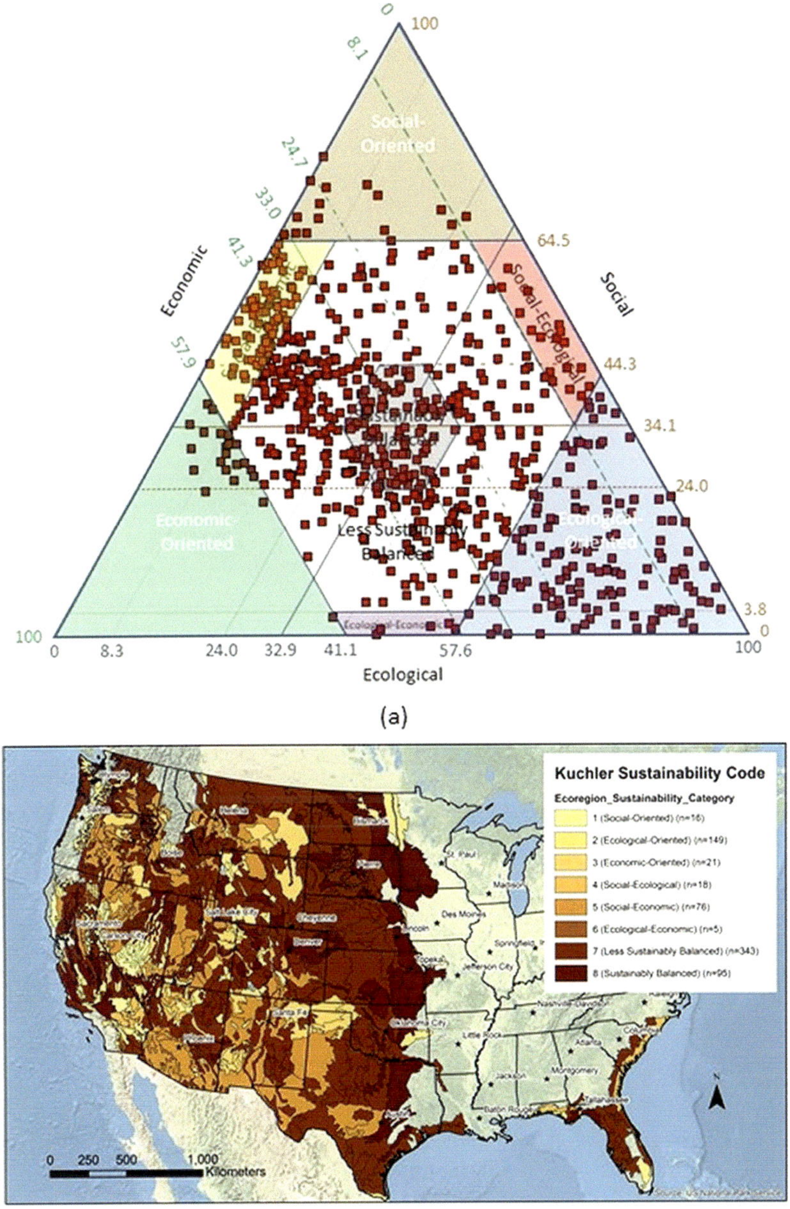

FIGURE 4.8
Sustainability Triangle (a) for KE Based on 2017 Data and the Map Distribution (b).

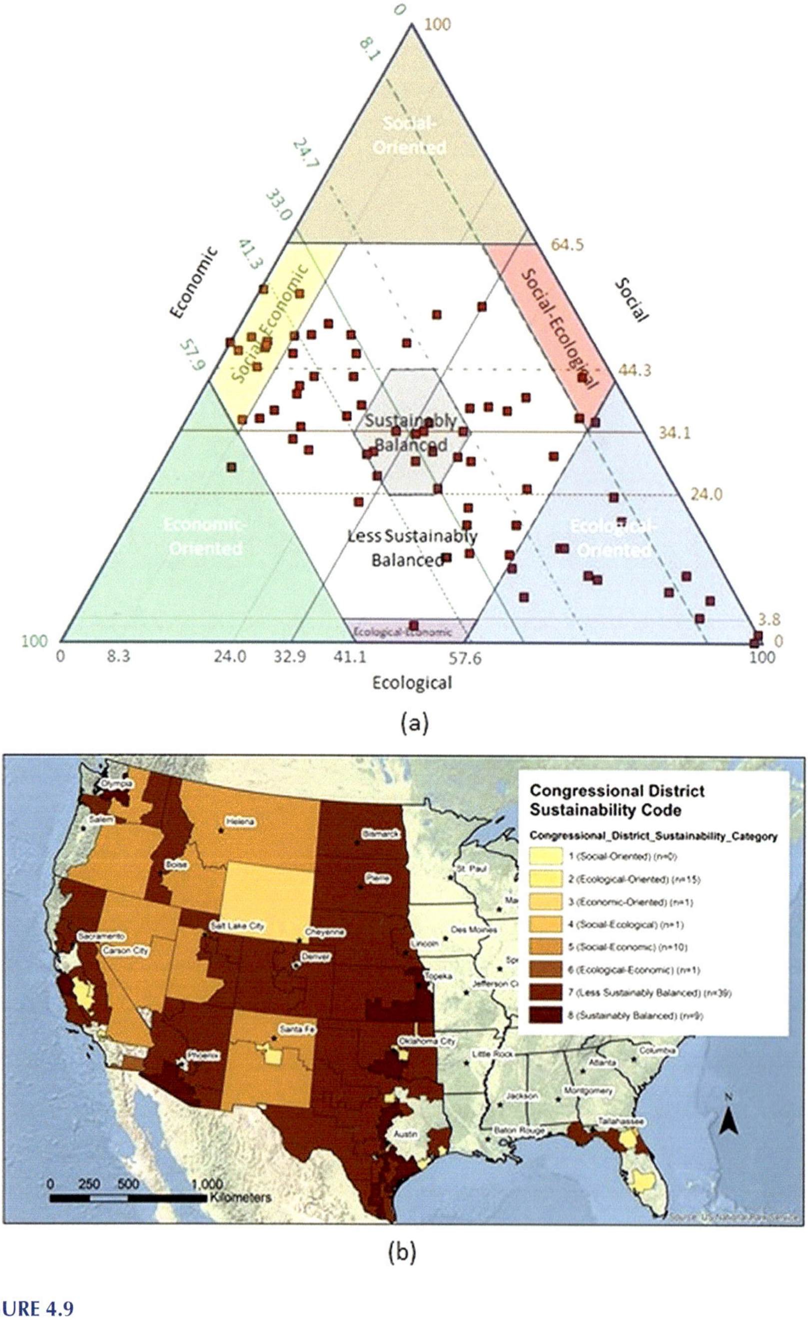

FIGURE 4.9
Sustainability Triangle (a) for CD Based on 2017 Data and the Map Distribution (b).

TABLE 4.3

Comparison of Ways to Score the Three Indicator Sets Using Hypothetical Data

	Social	Economic	Ecological	Total
Raw scores (0 to 100)	50	67	90	207
Adjusted to sum to 100	24	32	44	100
Weights	1	1	2	4
Raw scores times weight	50	67	180	297
Weighted scores adjusted	17	23	60	100

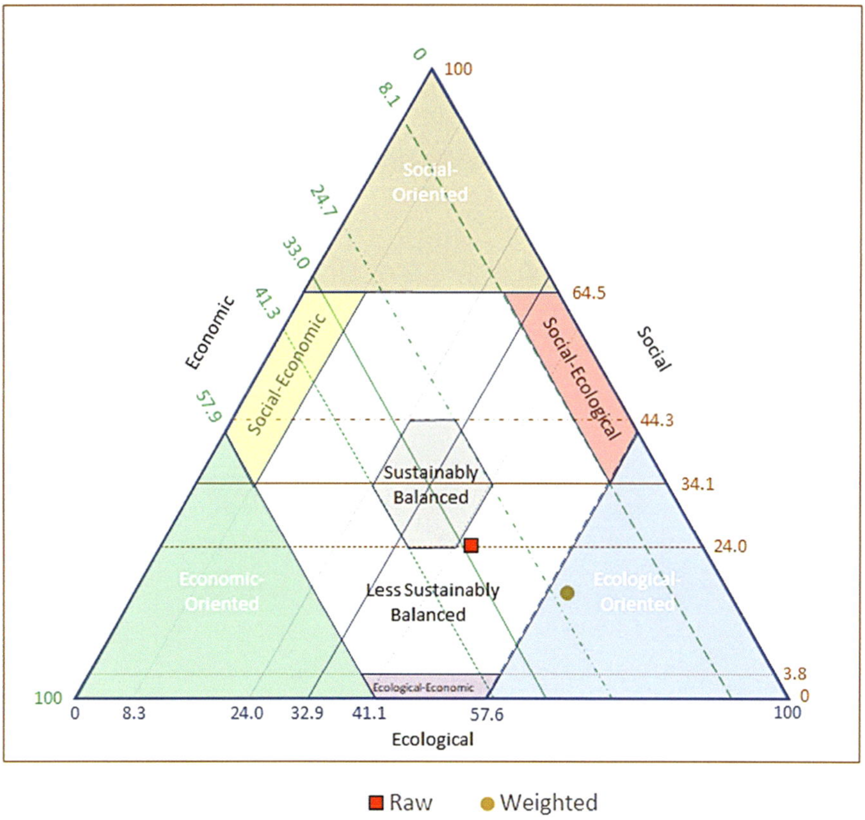

FIGURE 4.10
Effect of different weighting on assessing rangeland sustainability. The red box indicates unweighted scores. The green circle indicates when the ecological score is given twice the weight of the social and economic scores.

State-Level Sustainability

In addition to representing sustainability balance for all 690 rangeland counties, we can use the Sustainability Triangle on a state-by-state basis (Figure 4.11) for each state with rangeland counties. We can interpret the state figures in a variety of ways. For example, most states appear to have most of their rangeland counties in the sustainably balanced and less sustainably balanced portions and relatively centered. At the same time, the county scores appear to lean toward one side or the other (Figure 4.2b), such as Arizona, Idaho, Nevada, and Utah leaning toward social; Montana and

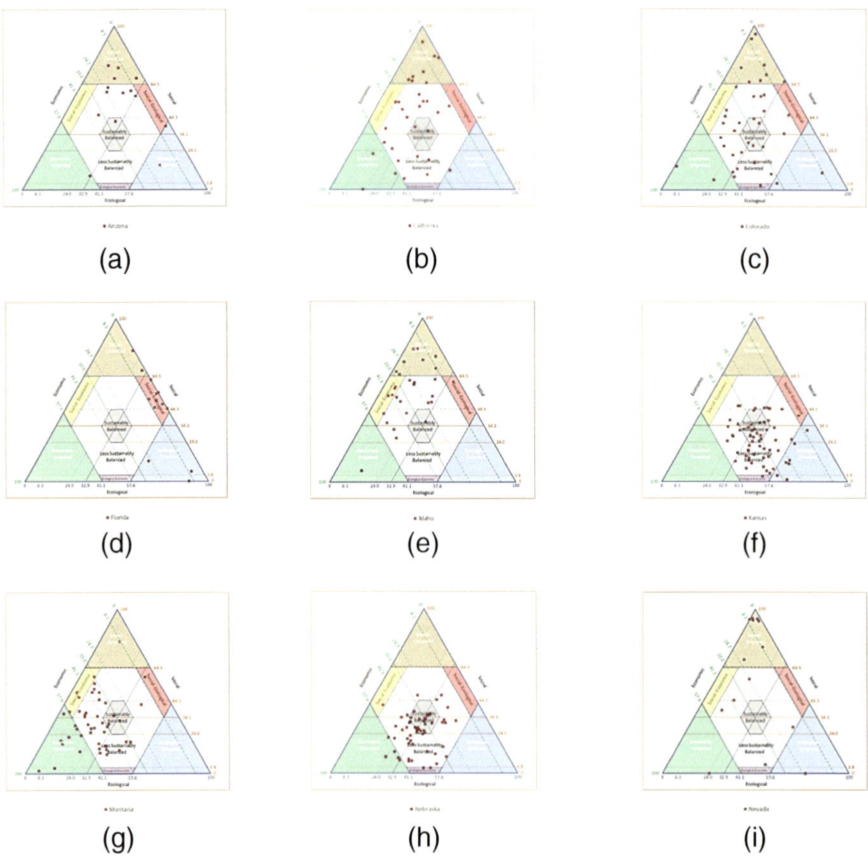

FIGURE 4.11
Individual state Sustainability Triangle plots. (a) Arizona, (b) California, (c) Colorado, (d) Florida, (e) Idaho, (f) Kansas, (g) Montana, (h) Nebraska, (i) Nevada, (j) New Mexico, (k) North Dakota, (l) Oklahoma, (m) Oregon, (n) South Dakota, (o) Texas, (p) Utah, (q) Washington, and (r) Wyoming. Red dots indicate counties.

(Continued)

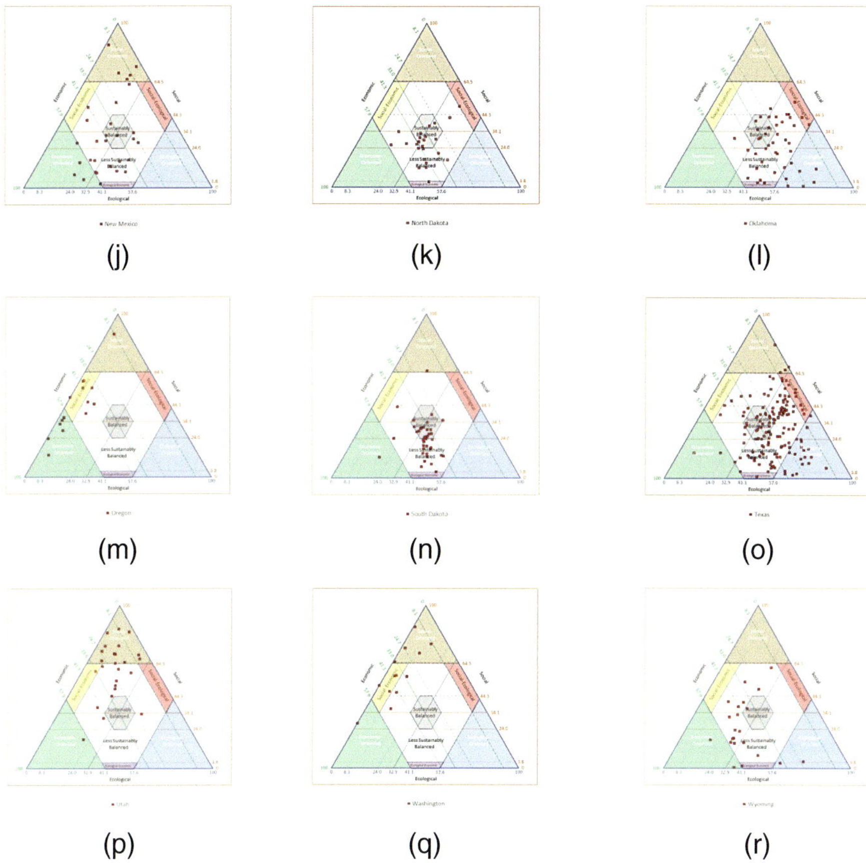

FIGURE 4.11 (Continued)

Wyoming toward economic; and Kansas and Oklahoma toward ecological. While all of the states have many of their counties in the two central areas, some of the remaining states have multiple counties in one of the unbalanced areas of the triangle and appear to be unsustainable in this one-year look.

Conclusion

This is just a way to depict rangeland sustainability, so there are caveats to its use. Changing weights to get the answer you want is a danger. Weights should be selected by an individual or a group before scoring begins.

The default of equal weights should be examined in every case as a baseline. The same sort of weighting can be done with individual indicators to come up with the economic, ecological, or social average score.

The Sustainability Triangle is presented here as a way to depict where a given community, state, region, or the nation lies. It quantifies sustainability based upon data collected for individual indicators, requiring that individual indicators be quantifiable rather than qualitative in nature. Our selection of breaking points as plus or minus 0.5 and 1.5 standard deviations from the mean was arbitrary, as were the names selected for each category (Table 4.2). More research is needed on where boundary lines are placed, what each section truly indicates, and how to interpret what it means to be within one of the sections.

Lastly, it is important to note that the raw data scores do not necessarily align with the re-standardized scores for use in the Sustainability Triangle. The scores in the triangle facilitate finding a combined balanced score and, hence, the triangle depicts balance among ecological, economic, and social data. A given county could rank low in all three categories and another high in all three; both counties would show up in the sustainably balanced area of the triangle. Clearly, the two counties would not both be considered equally sustainable. This conundrum will be explored in more detail in Chapter 5.

Literature Cited

Advocate, D., 2019. 12 ternary plot. https://www.youtube.com/watch?v=MqCvSa4fqBg and https://drive.google.com/file/d/1yi0D7E35UF-OkUjOTm2Jl-i2ww_dPrEY/view (accessed 20 July 2020).

Deutsch, M.E., 2011. The human footprint in the Bafing-Falémé Transboundary Protected Area: implications for biodiversity conservation and sustainable development. M.A. Thesis, International Studies/Environment and Natural Resources. University of Wyoming, Laramie, WY.

Leu, M., Hanser, S.E., Knick, S.T., 2008. The human footprint in the west: a large-scale analysis of anthropogenic impacts. *Ecological Applications* 18, 1119–1139.

Sanderson, E.W., Jaiteh, M., Levy, M.A., Redford, K.H., Wannebo, A.C., Woolmer, G., 2002. The human footprint and the last of the wild. *Bioscience* 52, 891–904.

[NRCS] Natural Resources Conservation Service, 2020. Soil triangle. https://www.nrcs.usda.gov/wps/portal/nrcs/detail/soils/survey/?cid=nrcs142p2_054167 (accessed 9 April 2020).

Woolmer, G., Trombulak, S.C., Ray, J.C., Doran, P.J., Anderson, M.G., Baldwin, R.F., Morgan, A., Sanderson, E.W., 2008. Rescaling the human footprint: a tool for conservation planning at an ecoregional scale. *Landscape and Urban Planning* 87, 42–53.

5

What the Data Show: Integrating Ecological, Social, and Economic Rangeland Sustainability

The Sustainability Triangle described in Chapter 4 is an attempt to integrate ecological, social, and economic indicators. While the triangle only shows balance among the three categories, it does put data into a common framework. We do view this as a limitation. Nevertheless, it puts the indicator information into a framework useful for dialogue and discussion. The Sustainable Rangeland Roundtable was created in part to define indicators that would be most appropriate to "define" sustainability. It was never intended to say whether data and their change over time were good or bad. As has been discussed previously, whether the data or change are good or bad or in the right direction depends on an individual's or group's personal perspectives or values. What is deemed good or bad may also change over time as values change. With that in mind, we want to be clear that we ranked each indicator from low to high based on our perception of what should be better. For example, a higher return on investment is deemed to be better than a lower return, and a lower incidence of poverty is deemed to be better than a higher incidence.

RANK SCORING

For all social, economic, and ecological indicators, raw data for each indicator were ranked among the 690 rangeland counties. Counties are categorized by how many standard deviations they are from the mean of all counties. All analysis and discussion in this chapter are based on counties, ecoregions, MLRAs, or congressional districts (CD) compared to all of the other counties, ecoregions, MLRAs, or CD, respectively, using the standard deviation categories.

While we would have liked to look at trends in the data, it was not feasible at this time to do so for the ecological data and some of the social and economic data. For the economic data, we did have access to much of it in 2002, 2007, and 2012, as well as 2017 that we used for this initial assessment. However,

DOI: 10.1201/9781003279747-5

NASS also changed some of the questions through time, deleting some and adding others. Hence there were two indicators we used in 2017 that would have only had data for 2012 and not earlier. Another example is the DCI (Economic Innovation Group, 2018). The DCI is built on five-year averages from the American Community Survey and is available going back to 2013 but not prior. So, while it would have been interesting to take a retrospective look at how the indicators changed through time, as depicted in the Integrated Social, Economic, and Ecologic Conceptual (ISEEC) framework as described by Fox et al., (2009) and shown in Figure 5.1, it is not feasible at this time.

In this ISEEC framework, ecosystem services are depicted as the nexus between the biophysical world and the human world. Ecosystem services are only imputed to have positive or negative value if people want or need them, or if they create adverse impacts that affect people. The boxed numbers indicate potential indicators that could be measured, and the arrows show directionality of the impacts. In the future, if additional data are collected that match the data used for this assessment, it may be possible to do a more complete sustainability assessment with trends over time.

We are not directly measuring sustainability of rangelands or communities presumed to be dependent upon them; what we actually measure are the indicators that give us some idea of the state of the systems (the SRR criteria). Ideally, we could explain why current conditions change through time as noted in Figure 5.1. We view this national assessment as the first step in the ISEEC process. That said, we can look at the 2017 data to start interpreting where rangeland nonurban counties rank in terms of sustainability indicators. Remember, we are not seeking to determine if the data show good or bad results; that is up to the individual reader to determine.

Defining Rangeland Counties

Rangelands vary by county. This includes ecological and human dimensions. We calculated the percent rangelands in each county using ArcMap and the National Vegetation Classification for all vegetation types considered to be rangeland (e.g., grassland, shrubland) (USNVC, 2019). We considered a county to be a rangeland county if it was greater than or equal to 25% rangeland. We further limited the counties we considered based on a population density of 200 people per square mile (United States Census Bureau, 2020) and classified those as nonurban (Figure 5.2). As shown, most of the counties have a population density below 5.2 people per square mile. Rangeland nonurban counties are shown in Figure 5.3.

In general, counties with a higher percent of rangeland have a lower population density. Figure 5.4 shows the relationship between rangeland area and population density. Data were combined from Figures 5.2 and 5.3. When

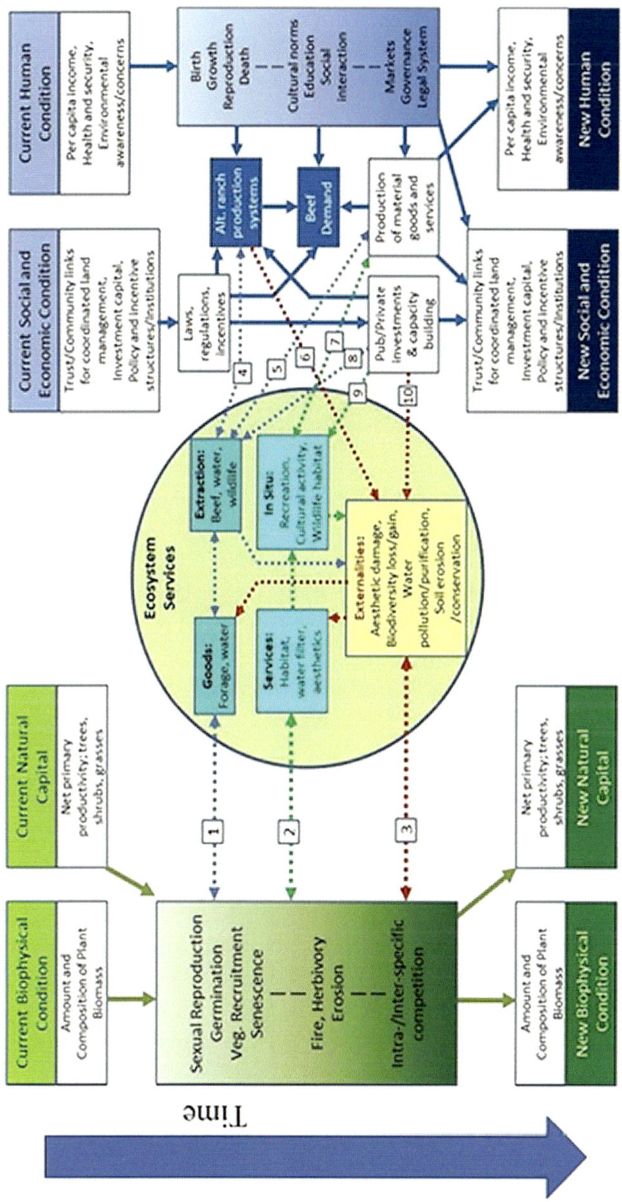

FIGURE 5.1
Integrated social, economic, and ecological conceptual framework showing indicators explaining change over time (Fox et al., 2009).

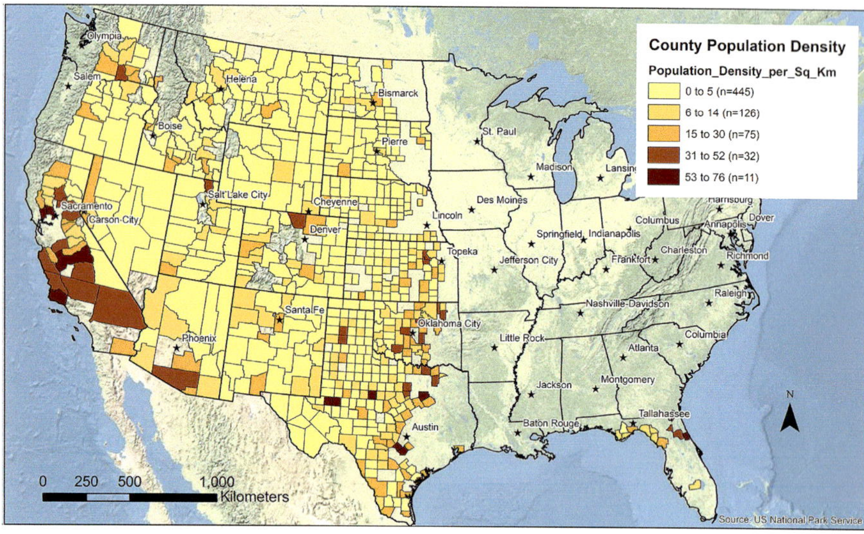

FIGURE 5.2
Population density below 200 people per square kilometer in counties with 25% rangeland or less.

counties had a low population density (0–2.2 people per square mile) and more rangeland, there were more counties. The trend is just the opposite for high population density counties.

Putting It All Together

As we combine the social, economic, and ecological indicators, we can do so at the county level, the MLRA level, the KE level, and the CD level. Additionally, the county-level indicators can be looked at on the state-by-state level. It is important to note that the data we are using is at the county level. We have aggregated it to MLRA, KE, CD, and state based on rangeland counties as we previously defined. After we aggregated to those larger scales, results were re-ranked and displayed.

So, what does all that mean? From the data we collected, it appears that the majority of counties have balanced social, economic, and ecological characteristics. There are some counties that achieve higher ranks on two of the three categories, and some that are ranked highly on only one dominant category. Is any of that good or bad? It depends on your personal perspective. Also, remember that the Sustainability Triangle only shows balance among ecological, social, and economic indicators. It does not allow us to conclude which indicators drive a similar result on the triangle.

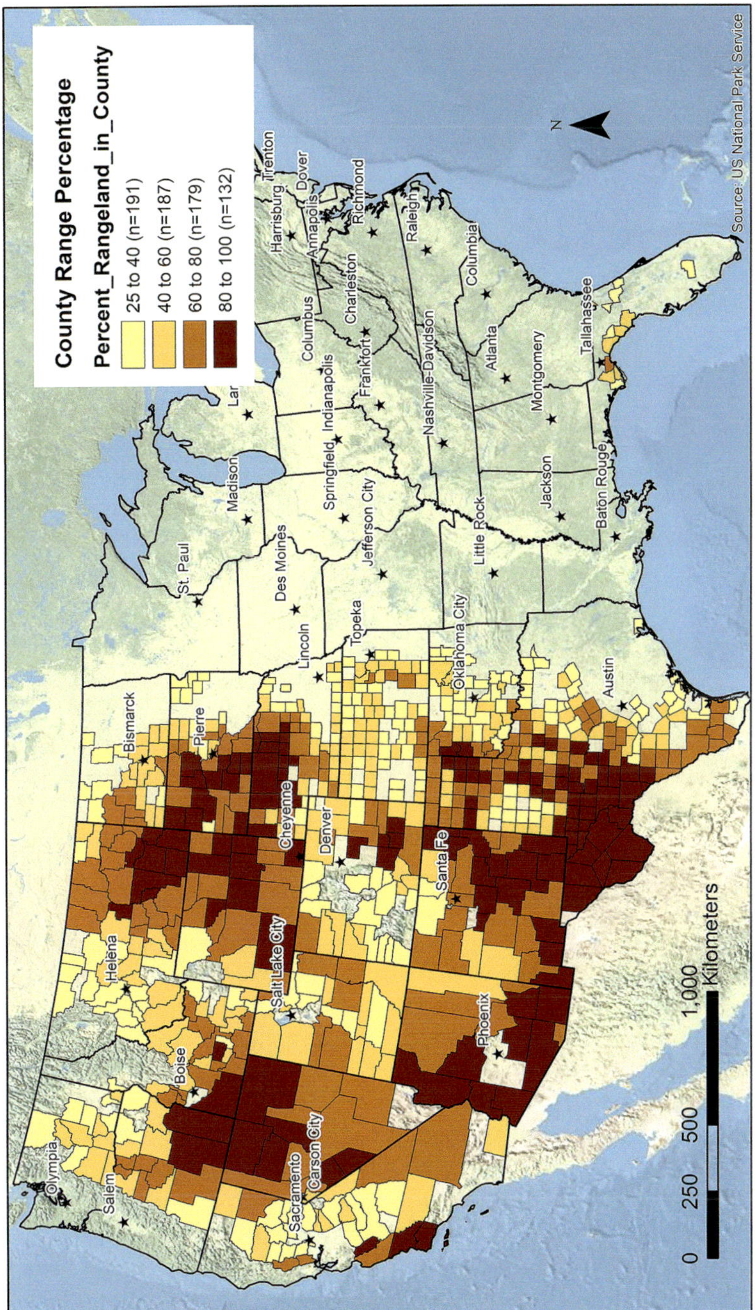

FIGURE 5.3
Percentage of rangeland in each rangeland nonurban county.

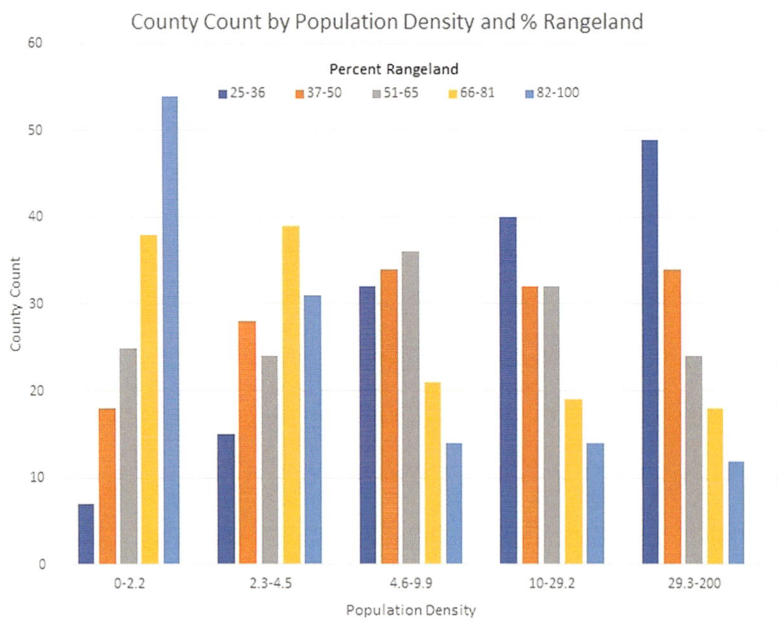

FIGURE 5.4
Count of counties based on quintiles of population density and percent of rangeland area.

County-Level and State-Level Indicators

Combining the ranks of county social, economic, and ecological indicators (Chapters 2 and 3) and the sustainability category (Chapter 4) and displaying on a map (Figure 5.5) shows counties that are ranked higher compared to other counties, and how balanced they are with respect to the Sustainability Triangle. The colors represent which counties fall within 0.5 or 1.5 standard deviations of the mean of the average indicator rankings, and the hatching/dots represent which category on the Sustainability Triangle the county falls within.

For examples of how Figure 5.5 can be interpreted, we can look at different parts of the map. The Northern Great Plains area appears to be ranked above average (darker colors) and falls within the Less Sustainably Balanced and Sustainably Balanced categories. While many counties in this region fall into these sustainability categories, not all do. The county south of Boise, Idaho (Owyhee county) is above average but falls within the Social-Economic category, meaning it would be low in ecological rankings. The Great Basin and Desert Southwest areas appear to be mostly below average (lighter colors) in total rankings and very diverse in what sustainability category each county falls within. Much of western Utah falls within the Social-Oriented category

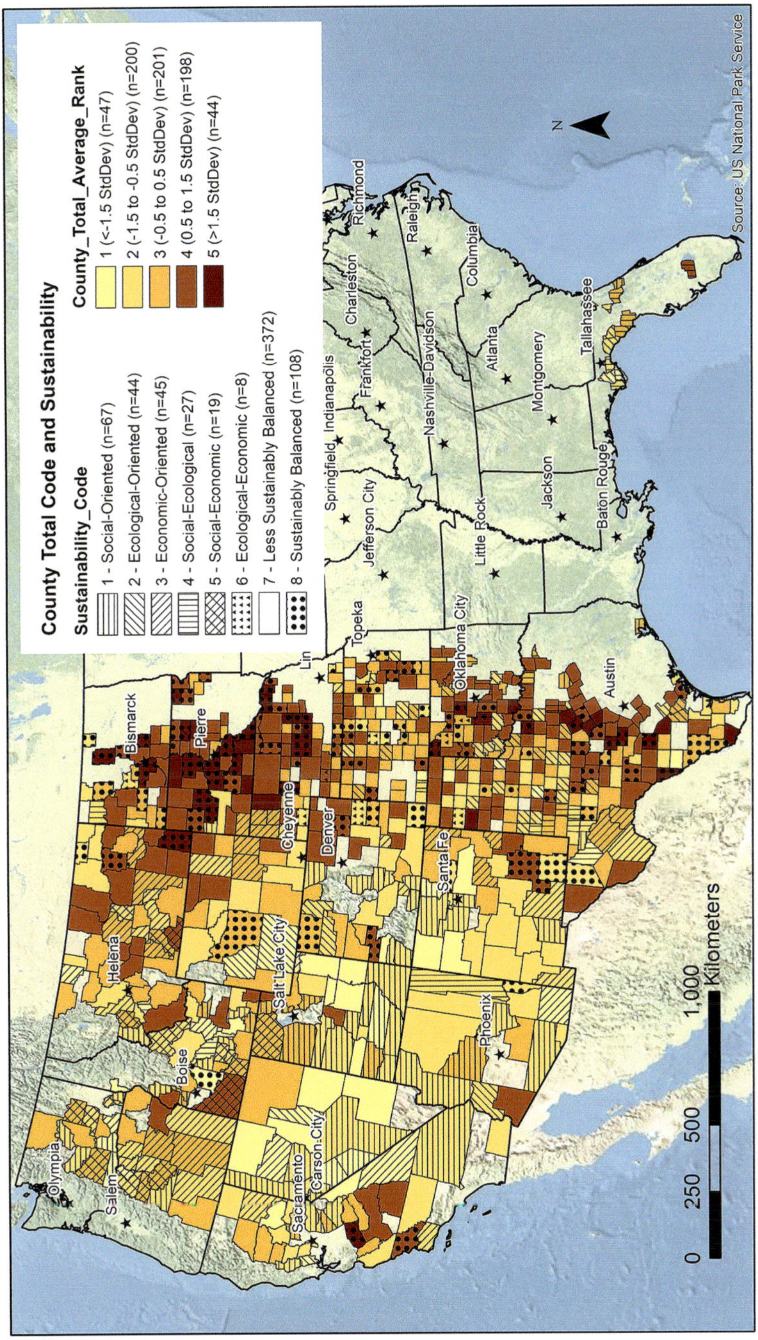

FIGURE 5.5
Representation of rangeland nonurban counties showing both the standard deviation from the mean range of overall indicator rankings (colors) and position on the Sustainability Triangle (hatching).

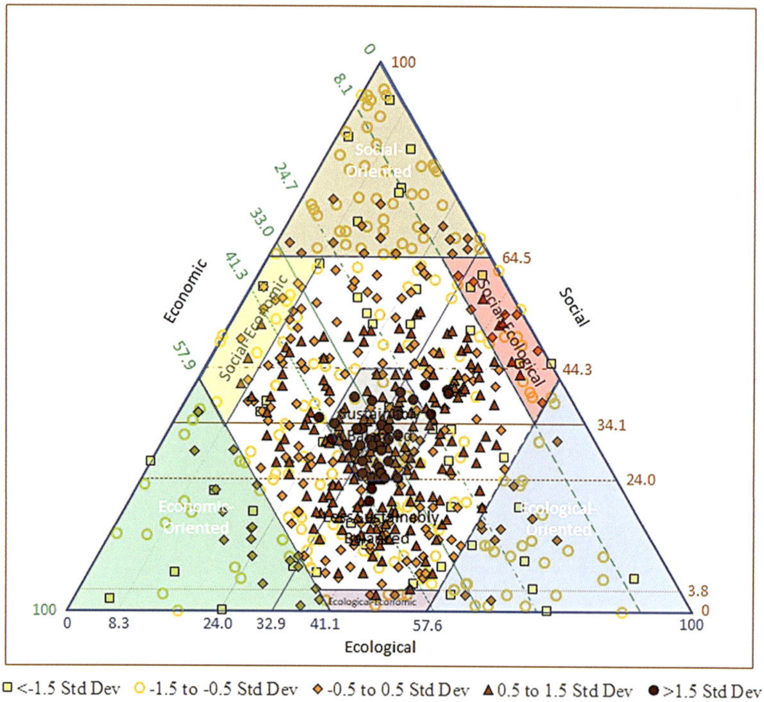

FIGURE 5.6
Cumulative rank category (1 to 5) scores for social, economic, and ecological indicators by state.

while eastern Nevada, northern Arizona, and western New Mexico appear to be primarily in the Less Sustainably Balanced category. This indicates that these latter regions balance the three sustainability categories but do so at a relatively low ranking (i.e., do less well than other counties).

We can also examine how the states rank among each other by taking an average of social, economic, and ecological indicator standard scores and comparing the totals (Figure 5.6). Nebraska and South Dakota score the highest totals of the three averages and Nevada scores the lowest. In examining Figure 5.5 colors, Nevada has a predominance of light-colored counties whereas Nebraska and South Dakota appear to have a predominance of dark-colored counties. To examine regional differences, we present three case studies in Chapter 6.

MLRAs, Ecoregions, and CD

As we examine the other spatial representation of the county-level data, it is imperative to remember that these were derived by allocating the county indicator data to each spatial area as a percentage of the county. In other words, if a county is 25% in one MLRA and 75% in a second MLRA, the miles of *impaired waters* are split 25:75 without regard to where those waters actually lie. Further, the whole MLRA is only based on rangeland county percentages and ignores nonrangeland counties within its boundaries. Those allocations were then re-ranked (comparing all MLRA units to the others) and new means and standard deviations calculated. The same process is applied to every indicator. This leads to a useful approximation for any spatial area beyond the county, but it reflects only our choice of how to disaggregate data for spatial units above the county level.

With that as a caveat, we can examine how different spatial scales look. Figure 5.7 shows the MLRA total average ranks and sustainability categories. Figure 5.8 shows the KE total average ranks and sustainability categories. Figure 5.9 shows CD total average ranks and sustainability categories. While there are differences among all spatial scales, results are similar, though with definite differences. For example, the Northern

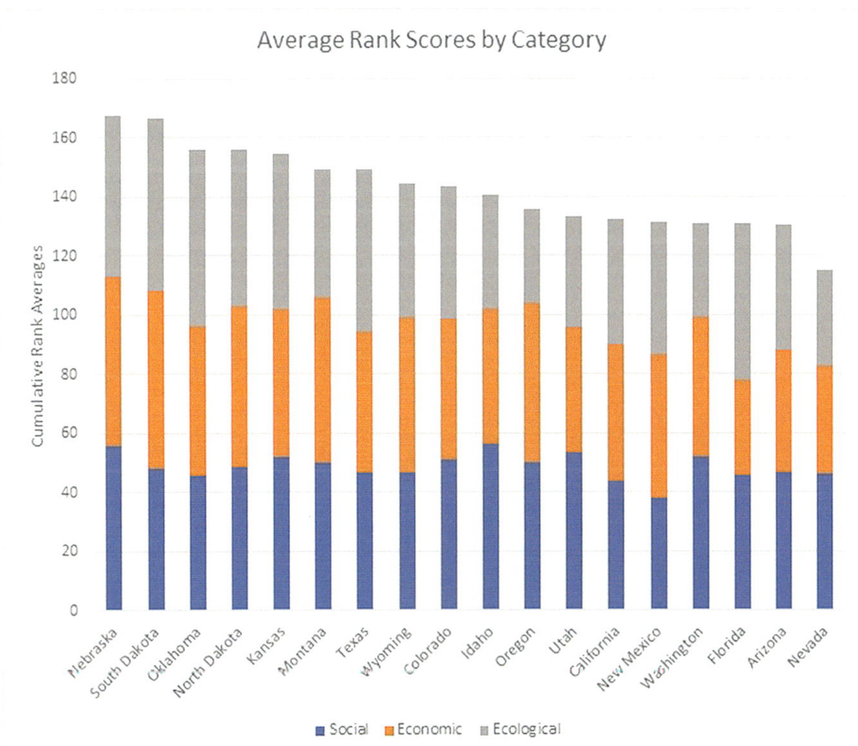

FIGURE 5.7
Representation of MLRAs showing both the standard deviation from the mean range of overall indicator rankings (colors) and position on the Sustainability Triangle (hatching).

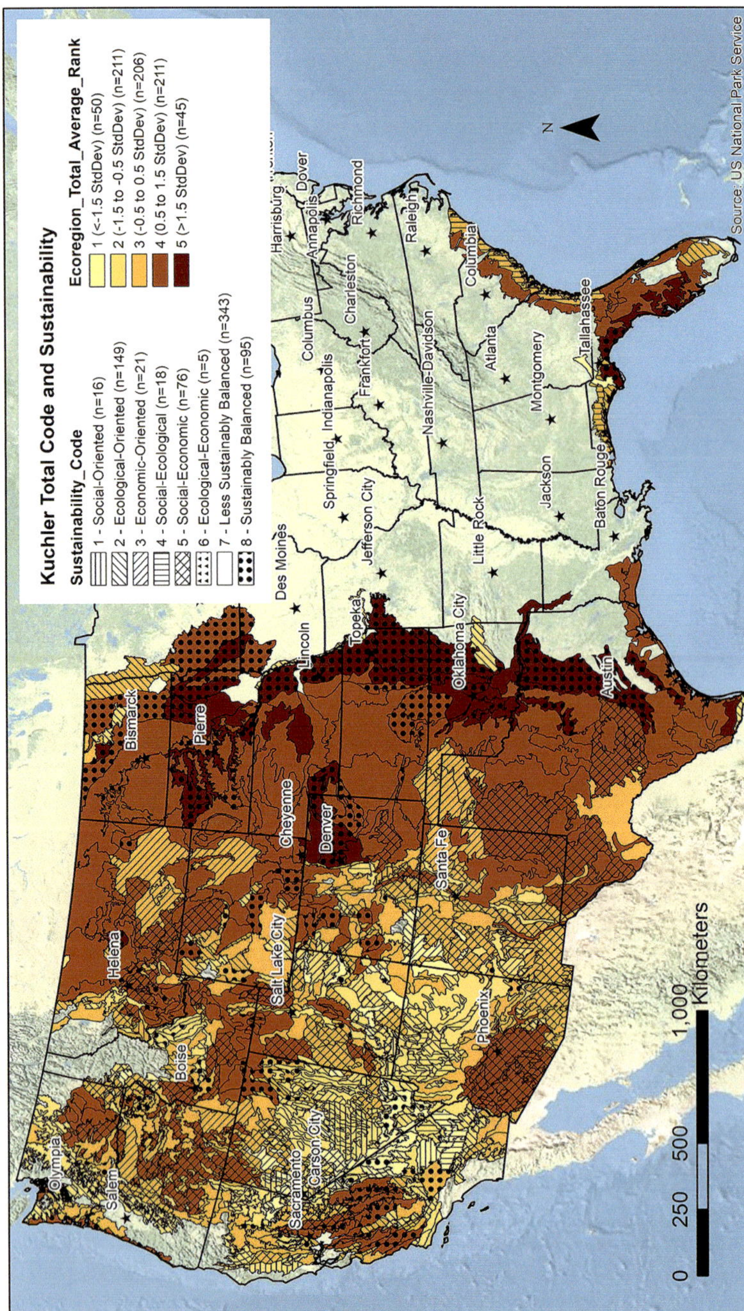

FIGURE 5.8
Representation of KE showing both the standard deviation from the mean range of overall indicator rankings (colors) and position on the Sustainability Triangle (hatching).

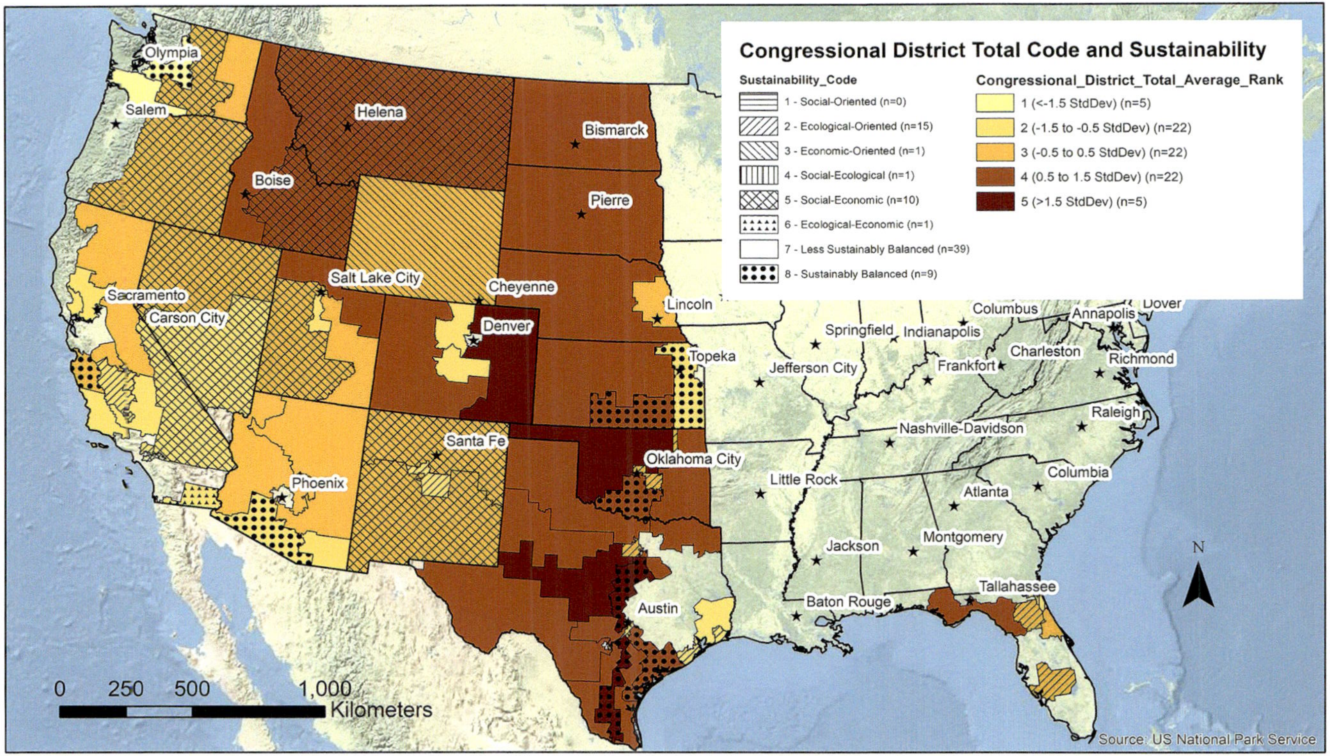

FIGURE 5.9
Representation of CD showing both the standard deviation from the mean range of overall indicator rankings (colors) and position on the Sustainability Triangle (hatching).

Great Plains counties are ranked higher (darker color) when compared to the other regions. That generally holds true for MLRAs and KEs but less so for CDs.

Literature Cited

Economic Innovation Group, 2018. *From great recession to great reshuffling: charting a decade of change across American communities*. Washington, DC. https://eig.org/wp-content/uploads/2018/10/2018-DCI.pdf (accessed 3 September 2020).

Fox, W.E., McCollum, D.W., Mitchell, J.E., Swanson, L.E., Kreuter, U.P., Tanaka, J.A., Evans, G.R., Heintz, H.T., 2009. An integrated social, economic, and ecologic conceptual (ISEEC) framework for considering rangeland sustainability. *Society and Natural Resources* 22, 593–606.

United States Census Bureau, 2020. County population totals: 2010–2019. https://www.census.gov/data/datasets/time-series/demo/popest/2010s-counties-total.html (accessed 15 December 2020).

United States National Vegetation Classification (USNVC), 2019. The United States National Vegetation Classification – your guide to inventorying natural and cultural plant communities. http://usnvc.org (accessed July 2019).

6

Case Studies

Case studies are presented to illustrate how the information presented in this book could be applied and interpreted. It is important to note that we will try not to make judgmental statements on what is good or bad, as that is up to the individual reader. We will endeavor to write about what the indicators mean when looked at as a whole, compared to other regions. We had considered evaluating the indicators on a county basis, but we concluded that it is more illustrative to talk about smaller regions that stand out. Counties within a region were selected based on their location and those that had at least 50% of their land area within the same MLRA (Figure 6.1). Each case study area has similarities among their counties, but they also have dissimilarities. We will attempt to interpret these, but it is really up to decision-makers within those areas to have the discussions.

RANK SCORING

For all social, economic, and ecological indicators, raw data for each indicator were ranked among the 690 rangeland counties. Counties are categorized by how many standard deviations they are from the mean of all counties. All analysis and discussion in this chapter are based on counties, ecoregions, MLRAs, or congressional districts (CD) compared to all of the other counties, ecoregions, MLRAs, or CD, respectively, using the standard deviation categories.

Figure 6.1 is based on county total standardized ranks that have been drawn from social, economic, and ecological indicators. We consider the Wasatch Front to be Box Elder, Cache, Juab, and Tooele Counties in Utah. The Central Great Plains area includes Stevens (Kansas), Cimarron and Texas (Oklahoma), and Sherman, Hansford, Ochiltree, and Moore (Texas) Counties. The Northern Great Plains area includes Garfield, McCone, Prairie, Rosebud, and Custer Counties in Montana. Figures 6.2 through 6.4 show the average standardized ranks for ecological, economic, and social indicators, respectively. We discuss each region independently in the following sections and refer back to these figures.

DOI: 10.1201/9781003279747-6

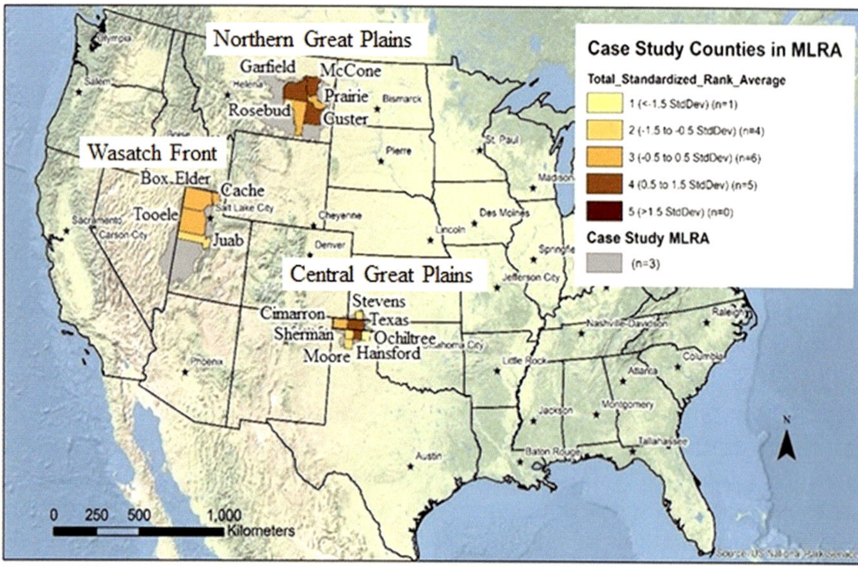

FIGURE 6.1
Case study counties shown within the three MLRAs. Colored counties are shown with the total indicator averages within the county.

Wasatch Front

This small region covers the counties surrounding Salt Lake City, Utah, including Box Elder, Cache, Juab, and Tooele Counties (Figure 6.1). For the counties in this case study, we estimate that 52% of the area is rangeland. It is 21% private and 44% federal grassland (NRCS, 2006). Livestock production is the primary use of the grasslands. Population density averages 12.4 people per square mile. Many of the residents of these counties commute into Salt Lake City for work and are therefore dependent upon it. Salt Lake County and the other counties usually considered the Wasatch Front are not part of the rangeland counties we considered in the assessment, yet those rangeland counties selected are vitally dependent upon this urban area. This area is an excellent example of rangeland areas existing in the presence of a large and growing urban area.

Wasatch: Ecological Indicators

The region is in the northern half of MLRA 28A – Great Salt Lake Area dominated by Utah juniper (*Juniperus osteosperma* (Torr.) Little), singleleaf pinyon (*Pinus monophylla* Torr. & Frém.), shadscale (*Atriplex confertifolia* (Torr. & Frém.)

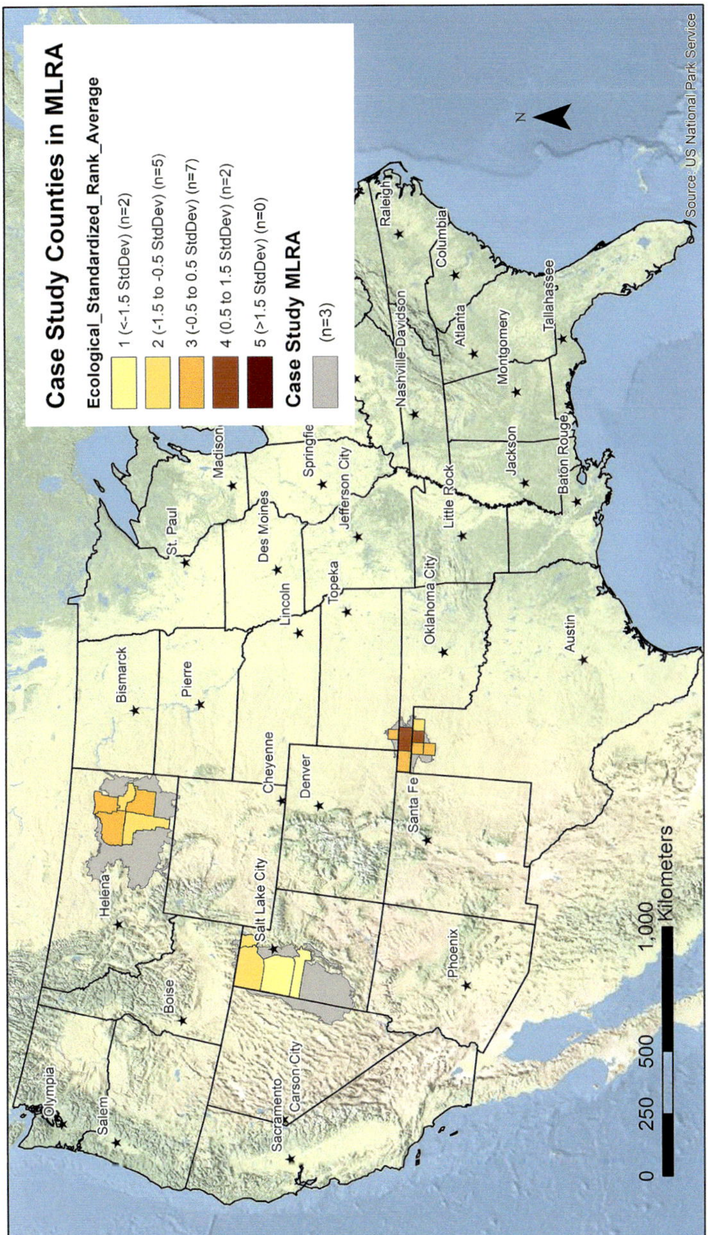

FIGURE 6.2
Ecological indicator standardized average ranks for case study counties.

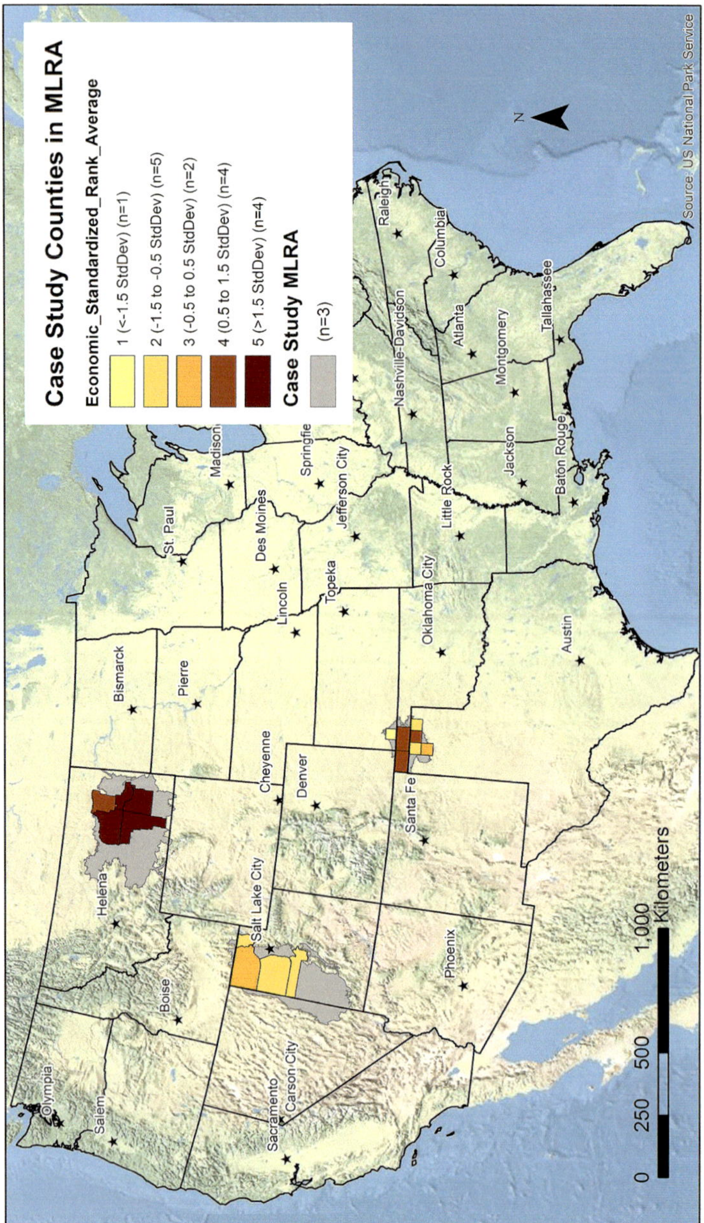

FIGURE 6.3
Economic indicator standardized average ranks for case study counties.

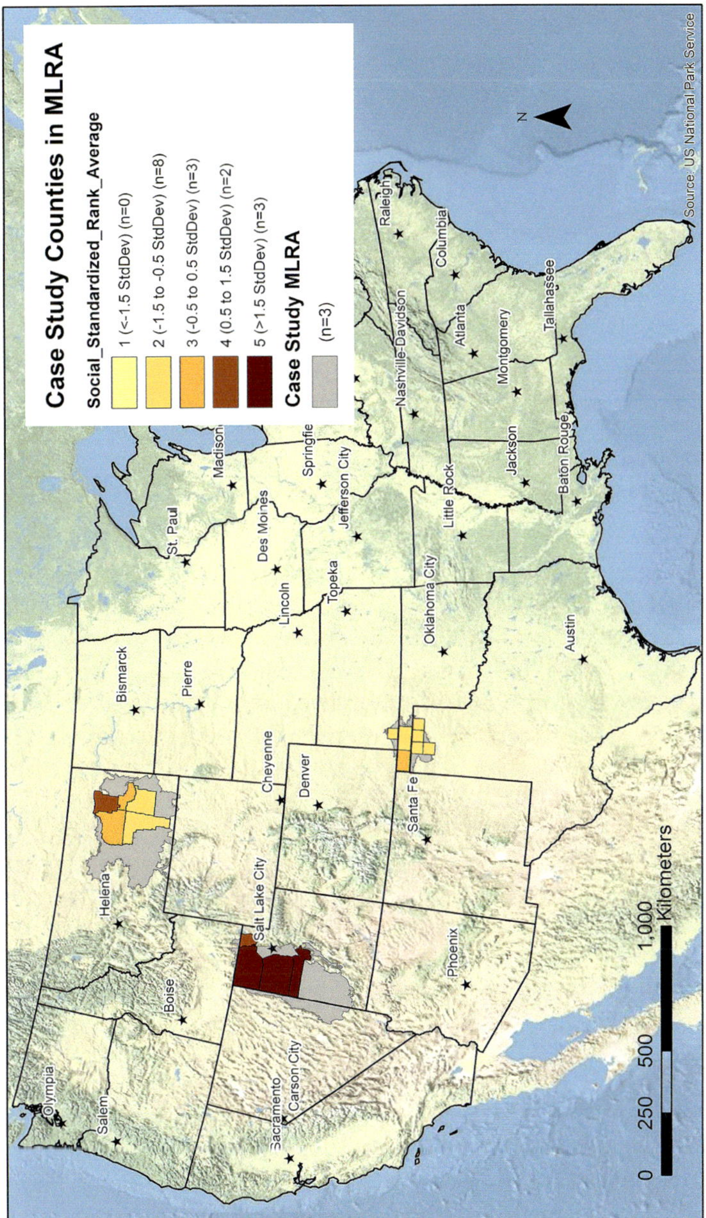

FIGURE 6.4
Social indicator standardized average ranks for case study counties.

TABLE 6.1

Ecological Indicator Standard Deviation Rank Categories[a] for Rangeland Counties in the *Wasatch Front Region*

Ecological Indicator	County				Indicator Average
	Box Elder	Cache	Juab	Tooele	
Beef Cattle, Sheep, and Meat Goat Numbers	5	3	4	4	**4.00**
Rangeland Acres	4	2	4	4	**3.50**
Impaired Waters	3	3	2	3	**2.75**
Rangeland Area by Vegetation Community	2	3	2	3	**2.50**
Phytomass	2	5	2	1	**2.50**
Annual Production	2	4	1	2	**2.25**
Fragmentation	2	2	2	2	**2.00**
Top Invasives	2	2	2	2	**2.00**
All Reported Invasives	2	1	2	2	**1.75**
County Average	**2.67**	**2.78**	**2.33**	**2.56**	
Wasatch Front Average	**2.58**				

[a] 1 = < −1.5 standard deviations, 2 = −1.5 to −0.5, 3 = −0.5 to 0.5, 4 = 0.5 to 1.5, 5 = > 1.5.

S. Watson), winterfat (*Krascheninnikovia arborescens* (Losinsk.) Czerep.), saltbush (*Atriplex canescens* (Pursh) Nutt.), black sagebrush (*Artemisia nova* A. Nelson), big sagebrush (*Artemisia tridentata* Nutt.), bluebunch wheatgrass (*Pseudoroegneria spicata* (Pursh) Á. Löve), Indian ricegrass (*Achnatherum hymenoides* (Roem. & Schult.) Barkworth), squirreltail (*Elymus elymoides* (Raf.) Swezey), bluegrasses (*Poa* L.), pickleweed (*Salicornia* L.), and seepweed (*Suaeda calceoliformis* (Hook.) Moq.) depending on location. Invasive annual grasses such as cheatgrass increasingly impact the MLRA (NRCS, 2006).

The ecological indicators by standard deviation categories are listed by county in Table 6.1. As shown in Figure 6.2, these counties are generally on the lower end compared to other rangeland counties as overall averages. The average of all ecological indicators is below three (within 0.5 standard deviations of the mean). For individual indicators, livestock number and rangeland acres rank high compared to other counties, and all reported invasive plant species rank low.

Wasatch: Economic Indicators

For the most part, the counties within the Wasatch Front are below average to average (standard deviation classes 2–3) as shown in Table 6.2 and Figure 6.3. Box Elder County seems to be doing economically better compared to the other three counties, especially in conservation easements, agricultural transfers, forage value, rangeland, and pasture acres and sales. Juab County appears to be ranked lower than the other counties in the region, especially

TABLE 6.2

Economic Indicator Standard Deviation Rank Categories[a] for Rangeland Counties in the Wasatch Front Region

	County				
Economic Indicator	**Box Elder**	**Cache**	**Juab**	**Tooele**	**Indicator Average**
Employment Diversity	3	4	3	4	**3.50**
Farm Recreation Income	3	4	3	4	**3.50**
Conservation Easements	4	4	2	2	**3.00**
Federal, State, and Local Transfers	4	3	3	2	**3.00**
Forage Value	4	2	2	4	**3.00**
Land Tenure	3	2	4	3	**3.00**
Land Use	3	2	3	4	**3.00**
Rate of Return on Investment	3	3	4	2	**3.00**
Rangeland and Pasture Acres	4	2	2	3	**2.75**
Sales, Livestock, and Business	4	3	2	2	**2.75**
Dependence on Livestock	3	3	3	2	**2.75**
Livestock per Ranch	3	2	4	2	**2.75**
Percent Agriculture and Recreation of GDP	3	3	2	3	**2.75**
Percent Family Farms	3	3	2	3	**2.75**
Livestock Sold	2	4	1	4	**2.75**
Ownership Patterns	3	3	2	2	**2.50**
Percent Beef Cattle Ranches	2	2	3	3	**2.50**
County Average	**3.18**	**2.88**	**2.65**	**2.88**	
Wasatch Front Average		**2.90**			

[a] 1 = < −1.5 standard deviations, 2 = −1.5 to −0.5, 3 = −0.5 to 0.5, 4 = 0.5 to 1.5, 5 = > 1.5.

in livestock sales, conservation easements, forage value, rangeland and pasture acres, importance of agriculture and recreation, and family farms. As a whole, the region is just below average (average rank category of 2.9).

Wasatch: Social Indicators

An overall description of this region from the social perspective is one of steady population growth and associated economic health. As shown in Table 6.3, three of the four counties in the region score well above the mean of all social indicators except the percent of the population over the age of 65. Note that for the DCI, all counties are above the mean, indicating they are prosperous relative to all other rangeland counties. In addition, the percent of the population under the age of 18 for each county is in the highest category above the mean, indicating a large percentage of children. Many areas

TABLE 6.3

Social Indicator Standard Deviation Rank Categories[a] for Rangeland Counties in the Wasatch Front Region

Social Indicator	County				Indicator Average
	Box Elder	Cache	Juab	Tooele	
% Population Under 18	5	5	5	5	**5.00**
Distressed Communities Index	5	4	4	5	**4.50**
Population Change 2013–17	4	4	5	5	**4.50**
Gini Coefficient	5	3	5	5	**4.25**
Adjusted Earnings	4	4	4	5	**4.25**
Unemployment Rate	4	4	3	3	**3.50**
% Population Over 65	2	1	2	1	**1.50**
Per Capita Total Transfers	1	1	2	1	**1.25**
County Average	**3.75**	**3.25**	**3.75**	**3.75**	
Wasatch Front Average	**3.63**				

[a] 1 = < −1.5 standard deviations, 2 = −1.5 to −0.5, 3 = −0.5 to 0.5, 4 = 0.5 to 1.5, 5 = > 1.5.

of the West are retirement destinations. While retirees may be moving to the Wasatch Front, the low rankings of this region for percentage of the population over the age of 65 and per capita total transfers indicate that population growth and economic health counterbalance the effects of retirees. Finally, note that all counties are in categories well above the mean for adjusted earnings. This measure indicates that these counties, relative to other rangeland counties, rely on the adjacent urban area for earnings flows.

Wasatch Front: Overall Conclusion

Rangeland area by vegetation community, *all invasives*, and *phytomass* are ranked low for ecological indicators, and all of the economic indicators are ranked fairly high. That being said, Table 6.4 shows that on average, ecological

TABLE 6.4

Average Rank by County for Social, Economic, and Ecological Indicators in the Wasatch Front Case Study Region

County	Ecological	Economic	Social
Box Elder	2.67	3.18	3.75
Cache	2.78	2.88	3.25
Juab	2.33	2.65	3.75
Tooele	2.56	2.88	3.75
Wasatch Front	**2.58**	**2.90**	**3.63**

TABLE 6.5

Average Standardized Rank by County for Social, Economic, and Ecological Indicators in the Wasatch Front Case Study Region

County	Ecological	Economic	Social
Box Elder	36.6	51.1	67.8
Cache	41.0	44.7	62.4
Juab	28.2	41.5	67.9
Tooele	32.1	44.4	67.8
Wasatch Front	**34.5**	**45.4**	**66.5**

indicators rank below average, economic indicators rank slightly below average, and social indicators rank well above average. Ecological indicators rank below both economic and social indicators for the four counties. In terms of the standardized score averages, the same message emerges (Table 6.5). Ecological indicators have standardized scores in the lower 40% of all rangeland counties, economic indicators roughly in the 40%–50% range, and social indicators in the upper third. This tells us that the region ranks better socially, about average economically, and lower ecologically.

If this region were to try to increase its social rankings relative to other rangeland counties, it might focus on federal transfers and people over 65. From an ecological viewpoint, it should focus attention on invasive species and fragmentation of the landscape. Individual counties likely would focus on different elements based on local conditions and potentials. In some cases, the indicators point to different things. For example, a county is unlikely to increase rangeland and pasture acres (increase its ranking), but a decrease in those over time may indicate why agriculture is becoming less important to the county economy or it may indicate a shift to higher intensity crop agriculture. A decrease in those acres may also be correlated with a decline in family farms that may also present social issues for communities.

As seen in Figure 6.5 on the Sustainability Triangle, counties in the case study region fall within different categories. Three of the counties would be considered social-oriented and one falls in the social-economic category. Three of the counties fall within 0.5 standard deviations from the mean, while one is below this average range. Again, this demonstrates that the Sustainability Triangle shows the level of balance among social, economic, and ecological indicator scores but not necessarily where the scores themselves lie. This aligns with the average rank scores (Table 6.4). If counties were to make decisions to move to a more sustainably balanced situation, they would need to make decisions to increase the importance of economic and/or ecological indicators. Those same counties may decide that they like where they sit in terms of sustainability and seek to improve further in the areas where they have some advantage.

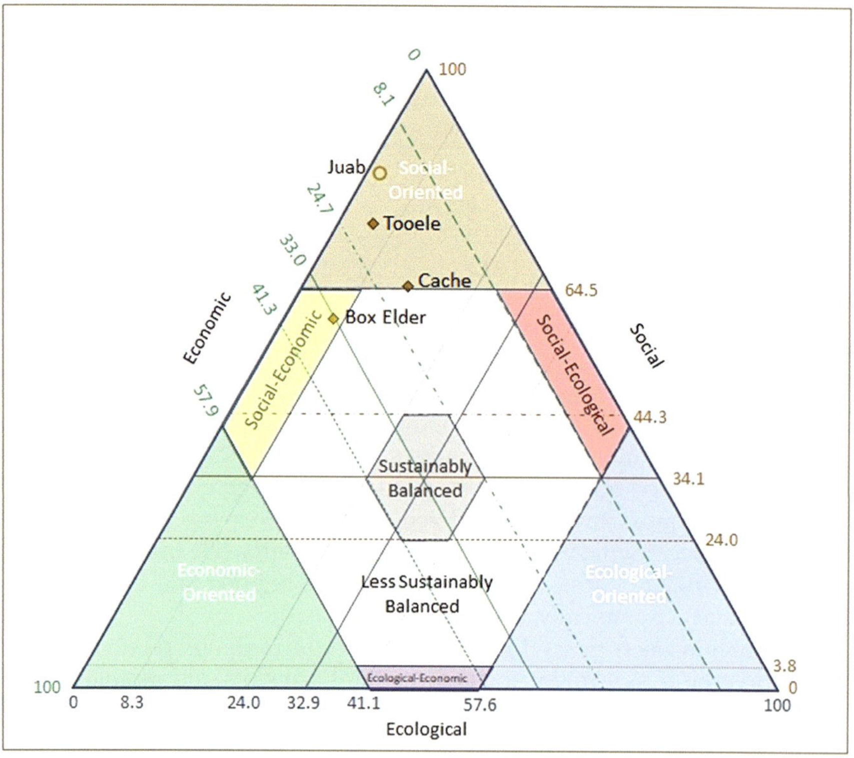

■ <-1.5 Std Dev ○ -1.5 to -0.5 Std Dev ◆ -0.5 to 0.5 Std Dev ▲ 0.5 to 1.5 Std Dev • >1.5 Std Dev

FIGURE 6.5
Sustainability Triangle for the Wasatch Front case study counties.

Central Great Plains

This region includes parts of southwest Kansas, the Oklahoma panhandle, and northern Texas. Stevens County, Kansas, Texas, and Cimarron Counties in Oklahoma, and Hansford, Moore, Ochiltree, and Sherman Counties in Texas. For the counties in this case study, we estimate that 58% of the area is rangeland. Ranches use the grasslands primarily for cattle, and beef and swine feedlots are important. Population density averages 3.2 people/km^2 and the region is characterized by declining populations.

Central Great Plains: Ecological Indicators

These counties were selected because they are within the MLRA 77A – Southern High Plains, Northern Part within the Central Great Plains Winter

Wheat and Range Region. The area is 32% private and 5% federal grassland dominated by blue grama (*Bouteloua gracilis* (Willd. ex Kunth) Lag. ex Griffiths) and buffalograss (*Bouteloua dactyloides* (Nutt.) J.T. Columbus) (NRCS, 2006).

The ecological indicators by standard deviation categories are shown by county in Table 6.6. As shown in Figure 6.2, these counties are generally on the lower end compared to other rangeland counties as overall averages. The average of all ecological indicators is slightly above three (within 0.5 standard deviations of the mean). For individual indicators, livestock numbers rank high compared to other counties, and annual production and rangeland acres rank low.

Central Great Plains: Economic Indicators

Counties within the Central Great Plains are about average as shown in Table 6.7 and Figure 6.3. Hansford County, Texas, seems to be doing economically better compared to the other counties in the region, ranking the highest in 6 of the 17 indicators. Stevens County, Kansas, appears to be ranked lower than the other counties in the region, ranking lowest in dependence on livestock for household income. As a whole, the region is just above average (average rank category of 3.03).

Central Great Plains: Social Indicators

The ranking of social indicators by standard deviation from the mean for this region relative to other rangeland counties is displayed in Table 6.8. Each county has an average rank for all social indicators that hovers around the mean, as measured by average ranks, slightly above and below 3.00. Taken as a region, the ranks are just below 3.00 with an overall average of 2.84. The region displays a pattern of population decline. All counties lost population between 2013 and 2017 and most are ranked below the mean for this indicator relative to other rangeland counties. Unemployment is low for the region. The variability of the DCI indicator suggests that these unemployment ranks might indicate tighter labor markets but do not necessarily translate into economic prosperity across the region. The high percentage of population under 18 years of age relative to all other rangeland counties and the low percentage of population over 65 is a common combination in this region. When counties are losing population, this might be construed as adults out-migrating and people with families retaining less prosperous jobs in the region. All of this contributes to the DCI showing two counties, Stevens County, Kansas, and Texas County, Oklahoma, as relatively prosperous, scoring above the mean for that indicator. On the other hand, Moore county, Oklahoma, is relatively distressed, scoring below the mean on the DCI.

TABLE 6.6

Ecological Indicator Standard Deviation Rank Categories[a] for Rangeland Counties in the Central Great Plains Region

	Kansas County	Oklahoma Counties		Texas Counties				
Ecological Indicator	**Stevens**	**Cimarron**	**Texas**	**Hansford**	**Moore**	**Ochiltree**	**Sherman**	**Indicator Average**
Beef Cattle, Sheep, and Meat Goat Numbers	3	3	4	1	4	2	4	**3.00**
Impaired Waters	2	4	2	3	4	3	2	**2.86**
Rangeland Area by Vegetation Community	2	3	5	2	3	3	2	**2.86**
Top Invasives	2	2	3	2	4	2	5	**2.86**
Fragmentation	2	3	4	2	3	4	1	**2.71**
All Invasives	2	2	4	2	3	3	3	**2.71**
Phytomass	2	3	4	2	3	4	1	**2.71**
Rangeland Acres	2	4	2	2	4	2	2	**2.57**
Annual Production	2	4	2	2	4	2	2	**2.57**
County Average	**2.89**	**3.00**	**3.56**	**3.33**	**3.11**	**2.78**	**3.11**	
Central Great Plains Average				**3.11**				

[a] 1 = < −1.5 standard deviations, 2 = −1.5 to −0.5, 3 = −0.5 to 0.5, 4 = 0.5 to 1.5, 5 = > 1.5.

TABLE 6.7

Economic Indicator Standard Deviation Rank Categories[a] for Rangeland Counties in the Central Great Plains Region

	Kansas County	Oklahoma Counties		Texas Counties				
Economic Indicator	**Stevens**	**Cimarron**	**Texas**	**Hansford**	**Moore**	**Ochiltree**	**Sherman**	**Indicator Average**
Livestock Sold	4	5	2	5	5	2	2	**3.57**
Ownership Patterns	3	4	4	5	2	3	4	**3.57**
Land Use	5	3	2	5	4	3	2	**3.43**
Sales, Livestock, and Business	3	2	4	4	5	5	1	**3.43**
Federal, State, and Local Transfers	4	3	2	5	3	2	4	**3.29**
Percent Agriculture and Recreation of GDP	4	4	1	5	5	2	2	**3.29**
Rate of Return on Investment	3	2	5	5	2	4	2	**3.29**
Percent Family Farms	4	4	1	4	4	1	4	**3.14**
Farm Recreation Income	3	2	4	3	3	3	4	**3.14**
Percent Beef Cattle Ranches	2	3	3	4	3	4	2	**3.00**
Land Tenure	2	2	5	2	4	5	1	**3.00**
Rangeland and Pasture Acres	3	2	4	3	4	3	1	**2.86**
Dependence on Livestock	1	2	4	3	4	4	2	**2.86**
Employment Diversity	4	1	2	4	5	2	1	**2.71**
Forage Value	2	2	3	3	4	2	3	**2.71**
Livestock per Ranch	2	2	3	3	3	4	2	**2.71**
Conservation Easements	2	1	3	4	4	3	1	**2.57**
County Average	**2.47**	**3.29**	**3.29**	**3.59**	**3.00**	**2.82**	**2.76**	
Central Great Plains Average				**3.03**				

[a] 1 = < −1.5 standard deviations, 2 = −1.5 to −0.5, 3 = −0.5 to 0.5, 4 = 0.5 to 1.5, 5 = > 1.5.

TABLE 6.8

Social Indicator Standard Deviation Rank Categories[a] for Rangeland Counties in the Central Great Plains Region

	Kansas County	Oklahoma Counties		Texas Counties				
Social Indicator	**Stevens**	**Cimarron**	**Texas**	**Hansford**	**Moore**	**Ochiltree**	**Sherman**	**Indicator Average**
Percent Population Age 18 or Younger	5	3	4	4	5	5	3	**4.14**
Unemployment Rate	4	5	4	4	4	3	4	**4.00**
Gini Coefficient	3	2	4	2	5	3	4	**3.29**
Distressed Communities Index	4	2	4	2	2	3	3	**2.86**
County Earnings Adjusted for Residence	3	3	4	2	2	2	4	**2.86**
Population Change, 2013 to 2017	2	1	2	3	3	1	3	**2.14**
Percent Population Age 65+	2	4	2	2	1	1	2	**2.00**
Per Capita Total Federal Transfers	1	3	1	2	1	1	1	**1.43**
County Average	**3.00**	**2.88**	**3.13**	**2.63**	**2.88**	**2.38**	**3.00**	
Central Great Plains Average				**2.84**				

[a] 1 = < −1.5 standard deviations, 2 = −1.5 to −0.5, 3 = −0.5 to 0.5, 4 = 0.5 to 1.5, 5 = > 1.5.

TABLE 6.9

Average Rank by County for Social, Ecological, and Economic Indicators in the Central Great Plains Case Study Region

County	Ecological	Economic	Social
Stevens	2.89	2.47	3.00
Cimarron	3.00	3.29	2.88
Texas	3.56	3.29	3.13
Hansford	3.33	3.59	2.63
Moore	3.11	3.00	2.88
Ochiltree	2.78	2.82	2.38
Sherman	3.11	2.76	3.00
Central Great Plains	**3.11**	**3.03**	**2.84**

Central Great Plains: Overall Conclusion

The lowest-ranked social indicator is *per capita total federal transfers*, the lowest-ranked ecological indicators are *rangeland acres* and *annual production*, and the lowest-ranked economic indicator is *conservation easements*. Table 6.9 shows that in total, social indicators are below average, ecological indicators are above average, and economic indicators are about average compared to all other rangeland counties. If this region wanted to focus on improvement, the lowest-ranked social indicators are related to having fewer retirement age people and low population growth. This is related to the region doing better on other social indicators, such as a higher young population and low unemployment. Ecologically, the lowest-ranked indicators are annual production and rangeland acres. The focus in these cases would be to manage the land to its potential and prevent land conversion in the future. On the economic front, there may be opportunities to improve employment opportunities and focus on conservation easements to protect land from conversion.

There may be indicators that some counties could prioritize for improvement over others. For example, four of the counties appear to have opportunities for recreation income. That would be a discussion for local landowners and policymakers. As shown in Figure 6.6, six of the case study counties fall in the less sustainably balanced category and one in the social-ecological category. In this case, even though the average ranking (Table 6.9) shows the overall ranking as more ecological, the individual counties are close to that direction but more balanced. Looking at the standardized scores for the counties (Table 6.10), it is more apparent that this region lags behind all rangeland counties in the social indicators and is about average for ecological and economic indicators. This demonstrates the conundrum of sustainability balance (Table 6.9) and relative

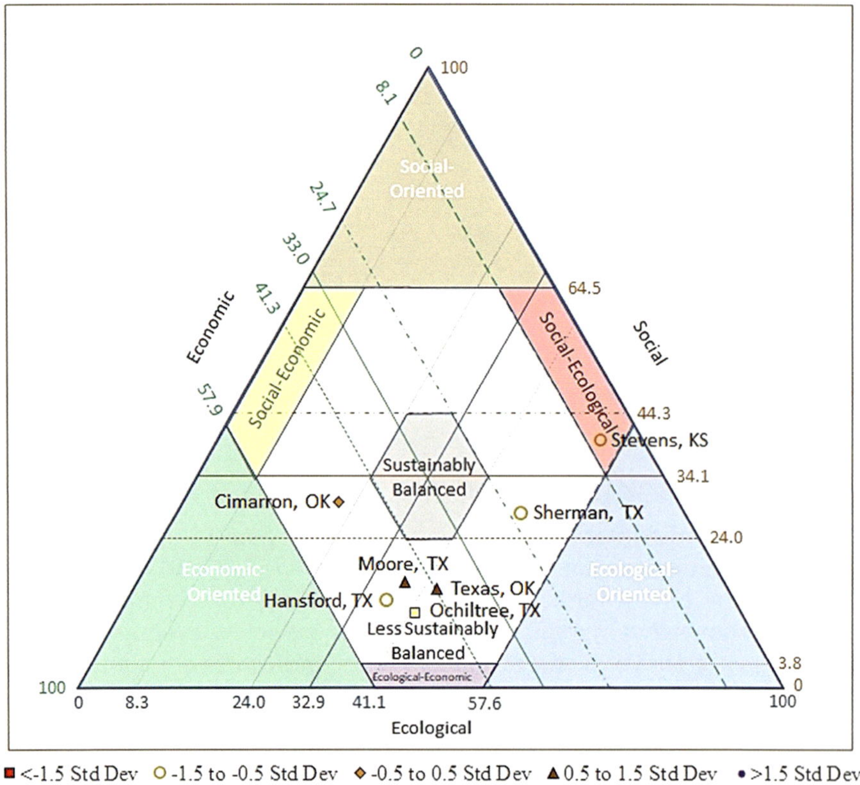

FIGURE 6.6
Sustainability Triangle for the Central Great Plains case study counties.

TABLE 6.10

Average Standardized Ranks for Social, Ecological, and Economic Indicators in the Central Great Plains Case Study Region

County	Ecological	Economic	Social
Stevens	44	5	33
Cimarron	36	80	49
Texas	86	82	31
Hansford	79	93	34
Moore	45	59	17
Ochiltree	27	30	8
Sherman	57	27	32
Central Great Plains	**51**	**51**	**29**

importance (Table 6.10), where the counties appear to be mostly in the less sustainably balanced category but score relatively low on the standardized scale.

Northern Great Plains

This region includes parts of eastern Montana and is comprised of Custer, Garfield, McCone, Prairie, and Rosebud counties. For the counties in this case study, we estimate that 52% of the area is rangeland. More than half of these rangelands are on ranches focused on livestock production. The largest town in the case study area is Miles City in Custer county. The Northern Cheyenne Reservation is in the southern part of Rosebud County. Population density averages 1.3 people per square mile.

Northern Great Plains: Ecological Indicators

These counties make up the northeastern portion of MLRA 58A – Northern Rolling High Plains, Northern Part within the western Great Plains Range and Irrigated Region. According to NRCS (2006), the area is 15% federal and

TABLE 6.11

Ecological Indicator Standard Deviation Rank Categories[a] for Rangeland Counties in the Central Great Plains Region

	County					
Ecological Indicator	**Custer**	**Garfield**	**McCone**	**Prairie**	**Rosebud**	**Indicator Average**
Rangeland Acres	4	5	4	4	5	**4.4**
Beef Cattle, Sheep, and Meat Goat Numbers	4	4	4	3	4	**3.8**
Impaired Waters	4	5	5	1	3	**3.6**
Rangeland Area by Vegetation Community	4	2	4	4	4	**3.6**
All Invasives	3	3	3	3	2	**2.8**
Fragmentation	2	2	3	2	2	**2.2**
Top Invasives	2	2	2	2	2	**2.0**
Annual Production	2	2	2	2	2	**2.0**
Phytomass	2	1	1	2	1	**1.4**
County Average	**3.00**	**2.89**	**3.11**	**2.56**	**2.78**	
Northern Great Plains			**2.87**			

[a] 1 = < −1.5 standard deviations, 2 = −1.5 to −0.5, 3 = −0.5 to 0.5, 4 = 0.5 to 1.5, 5 = > 1.5.

61% private grassland dominated by western wheatgrass (*Pascopyrum smithii* (Rydb.) Á. Löve), bluebunch wheatgrass, green needlegrass (*Nassella viridula* (Trin.) Barkworth), and needleandthread (*Hesperostipa comata* (Trin. & Rupr.) Barkworth). The remainder is in crops, forest, and urban development.

The ecological indicators by standard deviation categories are shown by county in Table 6.11. As shown in Figure 6.2, these counties are generally on the lower end compared to other rangeland counties as overall averages. The average of all ecological indicators is slightly below three (within 0.5 standard deviations of the mean). For individual indicators, rangeland acres and livestock numbers rank high compared to other counties, and annual production and phytomass rank low.

TABLE 6.12

Economic Indicator Standard Deviation Rank Categories[a] for Rangeland Counties in the Central Great Plains Region

Economic Indicator	County					Indicator Average
	Custer	Garfield	McCone	Prairie	Rosebud	
Rangeland and Pasture Acres	5	5	4	4	5	**4.6**
Livestock per Ranch	4	4	5	5	4	**4.4**
Percent Family Farms	4	5	4	3	4	**4.0**
Ownership Patterns	4	5	4	3	4	**4.0**
Federal, State, and Local Transfers	3	4	5	4	3	**3.8**
Land Tenure	4	4	4	3	4	**3.8**
Farm Recreation Income	4	4	3	3	4	**3.6**
Dependence on Livestock	4	4	2	4	4	**3.6**
Sales, Livestock, and Business	4	4	3	3	4	**3.6**
Percent Beef Cattle Ranches	4	4	2	4	4	**3.6**
Conservation Easements	4	3	2	5	3	**3.4**
Percent Agriculture and Recreation of GDP	4	4	3	3	3	**3.4**
Land Use	4	4	2	3	4	**3.4**
Rate of Return on Investment	3	3	3	4	3	**3.2**
Forage Value	2	2	3	4	3	**2.8**
Employment Diversity	4	2	2	2	2	**2.4**
Livestock Sold	2	2	2	2	2	**2.0**
County Average	**3.71**	**3.71**	**3.12**	**3.47**	**3.53**	
Northern Great Plains			**3.51**			

a 1 = < −1.5 standard deviations, 2 = −1.5 to −0.5, 3 = −0.5 to 0.5, 4 = 0.5 to 1.5, 5 = > 1.5.

TABLE 6.13

Social Indicator Standard Deviation Rank Categories[a] for Rangeland Counties in the Northern Great Plains Region

Social Indicator	County					Indicator Average
	Custer	Garfield	McCone	Prairie	Rosebud	
Unemployment Rate	4	4	5	3	2	**3.6**
Percent Population Age 65+	3	4	4	4	2	**3.4**
Gini Coefficient	4	2	3	4	3	**3.2**
Population Change, 2013 to 2017	3	4	3	2	3	**3.0**
Distressed Communities Index	4	3	3	2	2	**2.8**
Per Capita Total Federal Transfers	3	2	2	4	3	**2.8**
Percent Population Age 18 or Younger	2	2	3	3	4	**2.8**
County Earnings Adjusted for Residence	3	3	3	3	2	**2.8**
County Average	**3.25**	**3.00**	**3.25**	**3.13**	**2.63**	
Northern Great Plains			**3.05**			

[a] 1 = < −1.5 standard deviations, 2 = −1.5 to −0.5, 3 = −0.5 to 0.5, 4 = 0.5 to 1.5, 5 = > 1.5.

Northern Great Plains: Economic Indicators

Counties within the Northern Great Plains are above average as shown in Table 6.12 and Figure 6.3. Custer and Garfield Counties seem to be doing economically better, but all five counties are on the higher end. As a whole, the region is just above average (average rank category of 3.51).

Northern Great Plains: Social Indicators

The ranking of social indicators by standard deviation from the mean for this region relative to other rangeland counties is displayed in Table 6.13. County average ranks are mostly at the mean for all rangeland counties, except for Rosebud County, which ranks just below the average. The region has an overall average social indicator rank of 3.05 or almost right at the mean rank for rangeland counties generally.

Rosebud County differs persistently from the remainder of the region on most of the social indicators. For example, Rosebud has the highest rank for percent of population under 18 years of age and the highest rank for percent of population over 65 years of age. The county also has negative adjusted earnings, indicating that people are working in Rosebud County and living

TABLE 6.14

Average Rank by County for Social, Ecological, and Economic Indicators in the Northern Great Plains Case Study Region

County	Ecological	Economic	Social
Custer	3.00	3.71	3.25
Garfield	2.89	3.71	3.00
McCone	3.11	3.12	3.25
Prairie	2.56	3.47	3.13
Rosebud	2.78	3.53	2.63
Northern Great Plains	**2.87**	**3.51**	**3.05**

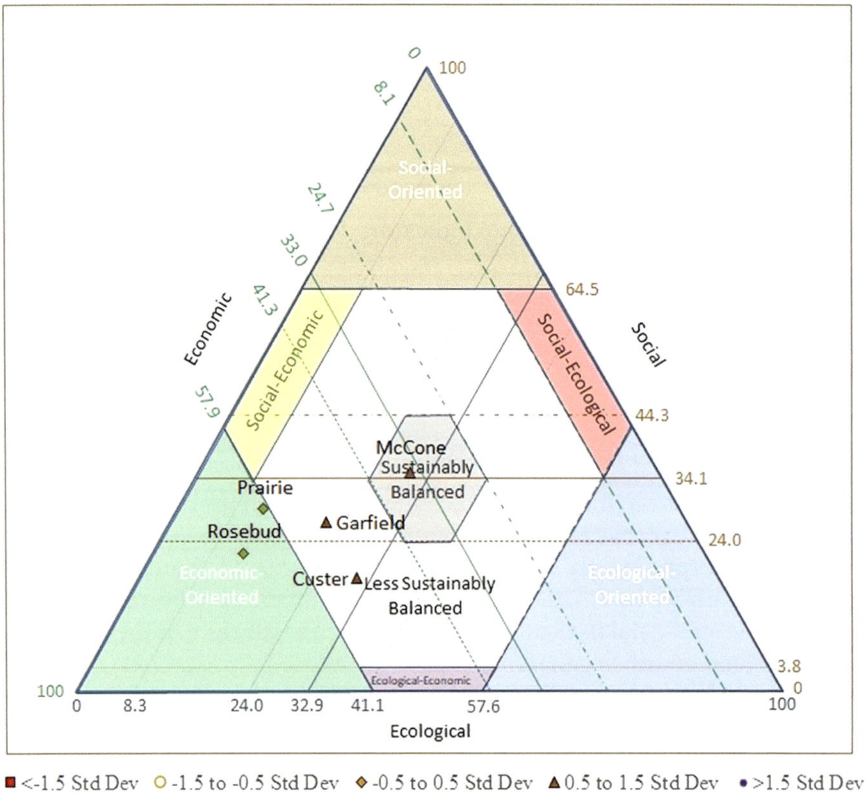

FIGURE 6.7
Sustainability Triangle for the Northern Great Plains case study counties.

TABLE 6.15

Average Standardized Rank by County for Social, Ecological, and Economic Indicators in the Northern Great Plains Case Study Region

County	Ecological	Economic	Social
Custer	59	99	35
Garfield	42	98	53
McCone	60	69	70
Prairie	20	94	47
Rosebud	18	94	32
Northern Great Plains	**40**	**91**	**47**

elsewhere. Finally, the DCI for Rosebud and Prairie Counties shows a ranking closer to distressed relative to other rangeland counties.

Northern Great Plains: Overall Conclusion

For the Northern Great Plains case study, relative to the rest of the rangeland counties, this region displays less emphasis on ecological considerations and more on economic (Table 6.14). Social considerations seem to be on par with many of the other rangeland counties nationwide. From a balanced perspective, the inhabitants of this region could improve on how they deal with the environment, especially in terms of major invasive species, which would be related to improving phytomass and production. Improving phytomass (standing vegetation) and annual production can be achieved through management changes, though it is somewhat limited on the natural potential defined by soils, precipitation, and temperatures. Relatively speaking, it will always rank lower than some of the southern counties just due to location. Hence, the rankings only give an indication and do not necessarily define where efforts should be placed. Within this region, there may be indicators that some counties could prioritize for improvement over others.

As shown in Figure 6.7, one county is in the sustainably balanced category, two are in the less sustainably balanced category, and two are in the economic-oriented category. The latter two counties (Prairie and Rosebud) show a lower ranking of their social and ecological indicators. In examining the standardized scores (Table 6.15) more differences can be observed.

Conclusion

We had no particular reason for selecting these three regions and the counties within them other than to show how the indicators can be used. They also illustrate that counties can fall in the same sustainability category within the Sustainability Triangle but have vastly different standardized scores. Another way of saying this is that a county that had social, economic, and ecological standardized scores of one and another with all three scores of 100 will show as being sustainably balanced (Figure 4.2a). If the higher standardized scores are taken as being better than a lower score, the first county would be balanced but not doing well compared to all of the other rangeland counties.

The discussion among involved stakeholders remains important in deciding what the indicators mean and where the opportunities lie for improvement. There are also relationships among several of the indicators that have been postulated in our discussion. We hypothesize those relationships and discuss them further in Chapter 7.

Lastly, the interpretation and local importance of any of the indicators depends on Criterion 5: legal, institutional, and economic frameworks for rangeland conservation and sustainable management. We did not include these indicators in this assessment for the variety of reasons outlined in Chapter 1. Nevertheless, they remain of utmost importance in deciding whether a policy can be changed to affect any of these indicators. For example, changing tax policy toward charitable donations may affect the size and amount of land put under conservation easements.

Literature Cited

[NRCS] Natural Resources Conservation Service, 2006. *Land resource regions and major land resource areas of the United States, the Caribbean, and the Pacific Basin*. U.S. Department of Agriculture Handbook 296, Washington, DC, p. 669+.

7

Key Elements Missing from Rangeland Sustainability Assessment: Future Research Needs

The Sustainable Rangelands Roundtable developed 5 sustainability criteria and 64 indicators based on what was known in the early 2000s. Specific potential data sources were identified, as were knowledge gaps. What was not known was whether the potential data sources were truly useful for a national assessment, especially over time. None of the indicators were actually populated with data through the SRR process. This book is the first attempt to do so in a meaningful way.

The most important thing we have discovered is that national-level data sets are messy and disaggregating them to states and counties is difficult. While the federal government collects a lot of data, over time, their methodologies change, questions are dropped and added, and the analysis of the data they have varies. For many indicators, national-level data are simply not collected, or, if they are, this may be at only one point in time.

In this chapter, we take a closer look at data needs both for the indicators we used for this national assessment and for those for which we had no known, usable, and acceptable data.

Ecological Data Needs

For the ecological data, very little is available at the national scale. Some of it must be pulled state by state; some are from citizen science, and some come from remote sensing information or spatial modeling. A lot of data is collected at different scales, using different methods. This makes attempts at aggregation suspect. While a number of programs and agencies regularly collect data for different purposes, very often they are only concerned with specific jurisdictions or have privacy concerns making aggregations challenging. With these factors in mind, we have grouped the indicators into categories for the purposes of discussing data and research needs—the focus of

DOI: 10.1201/9781003279747-7

this chapter. These categories include very likely, likely, not likely, and very unlikely, to describe the degree to which we think an indicator could realistically be quantified given more time or greater scrutiny of interpretation if another assessment were undertaken.

In a number of cases, the methods and analysis procedures are well codified (e.g., measuring soil aggregate stability), but the likelihood of collecting and analyzing soil aggregate stability to an extent useful for a national assessment is highly unlikely. In other cases, the indicator may be possible to evaluate but would require resources beyond those available for this assessment. For example, it may be possible to work with state game agencies to estimate the presence and density of wildlife functional groups on rangeland, but this would require a high degree of collaboration and a number of assumptions. In some cases, an indicator might have data available, but that data needs greater clarity to determine if it is in alignment with the intention of the indicator. Although the extent and condition of riparian systems may be possible to evaluate given some of the attributes available through the Environmental Protection Agency (EPA), the meaning of "condition" must be clarified.

It is worth noting that analysts may arrive at different conclusions about the feasibility of quantifying the ecological indicators discussed in this chapter. First, we address the ecological indicators not used in this assessment that have fragmented or inconsistent data at this time. This includes indicators for which data is gathered intermittently or at a spatial scale that is not useful here. Second, we discuss indicators for which little if any available data exists, is from sources previously unexplored, or for which such data is unlikely to be gathered.

Indicators Used in This Assessment

The following ecological indicators were used in this assessment, and we have no further suggestions.

Indicator 7: Percent of surface water on rangeland areas with significant deterioration of the chemical, physical, and biological properties from acceptable levels

Indicator 11: Extent of land area in rangeland

Indicator 14: Fragmentation of rangeland and rangeland plant communities

Indicator 21: Rangeland aboveground biomass

Indicator 22: Rangeland annual productivity

Indicator 18: Area of infestation and presence/absence of invasive and other non-native plant species of concern

Indicators with Partial or Fragmented Data

Indicator 4: Area and percent of rangeland with significant change in extent of bare ground

Time series of bare ground are now available that were not at the time we were acquiring data for the assessment. These data cover the western US rangelands and were derived from the Landsat Thematic Mapper satellite image time series from 1985 to 2020 (Robinson et al., 2019, Rigge et al., 2019). These data could be evaluated and significant changes in the bare ground could be quantified.

Indicator 6: Percent of water bodies in rangeland areas with significant changes in natural biotic assemblage composition

There are likely data available through the EPA that could be used to estimate this parameter, but it can be argued that the impaired water data set (EPA, 2020) used in the assessment could also be used to estimate this parameter. For example, the EPA website on water data and tools (EPA, 2021) has numerous sources that may have potential in future assessments but do not appear to be spatially explicit at a scale suitable for this assessment.

Indicator 9: Changes in the frequency and duration of surface no-flow periods in rangeland streams

Measurements of these changes can be made at any given stream gauge, but the answer is strongly dependent on the sites being measured. Only channel measures are normally made, but the valley fill capacity varies quite a bit depending on the valley geomorphology. Since we only measure the surface flow, it is never quite clear what the threshold of "zero" flow means in terms of actual down-valley water flux. Combining information from multiple stations would need to be done thoughtfully. Perhaps even more relevant is which places tend to go dry or stay wet and what is the water quality of the remaining locales. Measurements are very rare (Charlie Luce, personal communication, 11 January 2021). Perhaps a tool such as the FLOwPER (FLOW PERmanence) will aid the mapping of the presence of flow in streams (USGS, 2020).

Indicator 12: Rangeland area by plant community

It is possible to evaluate this indicator using two general approaches. First, maps of US ecological systems (Comer et al., 2003) through the Landfire Project (Rollins, 2009) can be used to quantify the vegetation types. Second, national plot-level inventories include the Assessment Inventory and Monitoring (AIM; administered by the US Department of Interior Bureau of

Land Management), National Resources Inventory (NRI), (administered by the US Department of Agriculture, Natural Resources Conservation Service), and to a more limited extent on rangelands the Forest Inventory and Analysis (FIA; administered by the US Department of Agriculture, Forest Service). The main issues here are

1) AIM data only cover Bureau of Land Management land;
2) NRI data only cover private land but, due to privacy concerns, these cannot be aggregated to county-level estimates;
3) FIA data only cover lands managed by the Forest Service and do not regularly sample rangelands as the main focus is forestlands.

Indicator 13: Number and extent of wetlands

The National Wetlands Inventory produces a periodic national assessment, but the baseline data covering the nation still emanates from the original information discovered in the 1980s. Subsequent changes and additions have been conducted in a piecemeal manner without consistent guidelines.

Indicator 15: Density of roads and human structures

Two data sets now available to populate this indicator include the 2018 Microsoft national buildings database (Microsoft, 2021) for buildings, the Landscan Nighttime Light Detection (Oak Ridge National Laboratory, 2021) for human structures or density, and Here Maps 2020 (Here, 2021) for roads. It seems likely that these data could be used to estimate density of roads and human structures.

Indicator 16: Integrity in natural fire regimes on rangeland

Robust debate exists regarding natural fire regimes. For example, it is difficult to know with certainty what constitutes a natural fire regime in many rangeland ecosystems, such as sagebrush steppe. However, it may be possible to still evaluate this indicator by evaluating trends in burned areas in especially sensitive areas such as the Sonoran Desert and sagebrush steppe ecosystems.

Indicator 17: Extent and condition of riparian systems

Surveying even portions of the immense riparian systems in a suitable manner to enable national assessment is not feasible in the foreseeable future. Perhaps in the future, advances in unmanned aircraft systems may enable a sampling framework to augment ground-based reconnaissance, although the costs for this type of protocol are presently prohibitive.

Indicator 19: Number and distribution of species and communities of concern

It may be possible to evaluate this indicator, but it would require considerable collaboration among federal, state, and local agencies to identify the abundance of species and communities of concern. Coordination through the International Union for Conservation of Nature framework and closely aligned organizations such as NatureServe (2021) would be a valuable starting point.

Indicator 20: Population status and geographic range of rangeland-dependent species

It may be possible to evaluate this indicator, but it would require considerable collaboration among federal, state, and local agencies to identify the population status and geographic range of rangeland-dependent species. If the focus could remain on a few species, such as the iconic sage-grouse (*Centrocercus urophasianus*), the likelihood for success would increase since the distribution and population dynamics are relatively well quantified. Engaging state fish and game agencies would be a critical step to success.

Indicator 23: Percent of available rangeland grazed by livestock

The number of livestock and the amount of rangeland and rangeland production can be quantified for each county using the data described in this report. However, the number of livestock and duration of use are unknown. It may be possible, through time, to develop some regional guidelines that could be applied to make assumptions about these parameters.

Indicator 24: Number of domestic livestock on rangeland

The number of beef cattle, sheep, and meat goats is available from the Census of Agriculture every five years. What is not known is how many of those are on rangeland. We assume that most of these animals in rangeland counties spend some part of the year on rangelands, but we do not know how many or for how long.

Indicator 25: Presence and density of wildlife functional groups on rangeland

It may be possible to evaluate this indicator, but it would require considerable collaboration among federal, state, and local agencies to identify the presence and density of wildlife functional groups. If the focus could remain on a few species, such as the iconic sage-grouse, the likelihood of success would increase since the distribution and population dynamics are relatively well quantified. Engaging state fish and game agencies would be a critical step to success.

Indicators with No Available Data

Creating a sampling framework to enable county-level estimates of this parameter is unlikely in the near future. Many of these indicators have research in their infancy or in limited locations. It may be many years before data are identified at the county scale in any meaningful way.

Indicator 1: Area and percent of rangeland soils with significantly diminished organic matter and/or high carbon:nitrogen ratio
Indicator 2: Extent of rangelands with changes in soil aggregate stability
Indicator 3: Assessment of microbial activity in rangeland soils
Indicator 5: Area and percent of rangeland with accelerated soil erosion by water or wind

While there may be such databases for cropland, it is unlikely to know much about these at large scales for rangelands.

Indicator 8: Changes in groundwater systems

This is a difficult parameter to estimate even when site-level parameters are known. As a result, it is unlikely that consistent data describing this parameter will be available in the near future.

Indicator 10: Percent of stream length in rangeland catchments in which stream channel geometry significantly deviates from the natural channel geometry

We do not have good information about channel morphology for all streams across the rangeland domain let alone what the "natural" geometry is. As a result, it is unlikely that this parameter can be estimated in the near future. Hafen et al. (2020) discussed some of the problems with estimating these sorts of parameters, as climate regimes have changed since many of our base data describing hydrography were created.

Indicator 26: Annual removal of native hay and non-forage plant materials, landscaping materials, edible and medicinal plants, and wood products

It is very unlikely that this indicator can be reliably quantified across US rangelands.

Social Data Needs

Social indicators are derived from SRR Criterion 4: maintenance and enhancement of multiple economic and social benefits to current and future generations. Most of the social data is taken from the decennial US Census. Between census years, options include the Current Population Survey (CPS) and the American Community Survey (ACS). These products are different in their samples and methods, and we chose to use the ACS because it is sampled in such a way that it can account for smaller area populations by using five-year samples (US Census Bureau, 2016; Every Texan, 2020). In addition, our choice to utilize the DCI (Economic Innovation Group, 2018) presented an opportunity to use the same source for other indicators to the extent possible. At a minimum, using the same data source offers many of our indicators the same sampling methodology (and resulting error rates).

Bureau of Labor Statistics gathers unemployment data monthly and annually for counties. The Department of Commerce, Bureau of Economic Analysis provides earnings adjusted for place data annually. These two series are available for trend analysis and could easily be examined if the other indicators were available in the same time-series format.

This chapter addresses data and research needs, and here we look at social indicators. In that context, we can look at each of the original indicators from SRR Criterion 4 in three ways. First, we discuss some, though not all, of the social indicators used in this assessment that can readily be expanded with available data and insight from existing research. Second, we examine indicators not used in this assessment that have fragmented or inconsistent data at this time. This includes indicators for which data are gathered intermittently or at a spatial scale that is not useful here. Finally, we discuss indicators for which little if any available data exists, is from sources previously unexplored, or for which such data is unlikely to be gathered.

Indicators Used in This Assessment

Indicator 37: Poverty rate (children)

The general poverty rate for a county is utilized in the DCI used in this analysis. Child poverty could be split from that overall rate to allow a focus on the structure of poverty in rangeland counties specifically. This would be relatively easy to accomplish with the available data. Further analysis would focus on such topics as the difference between child poverty in rangeland counties that are very rural and those with larger, urban areas within their boundaries.

Indicator 41: Federal transfers by categories (individual, infrastructure, agriculture, etc.)

We are currently using both a per capita total federal transfers indicator (Indicator 41) and federal, state, and local transfer to ranches (Indicator 41) for this analysis. County-level data for Indicator 41 can be disaggregated easily into more detailed categories from Bureau of Economic Analysis data. General examples could be income support for families and individuals compared to Social Security payments to the elderly. This allows an examination of the relative role that different categories of transfers play in income streams available to communities in rangeland counties.

Indicator 48: Employment, unemployment, underemployment, and discouraged workers by industrial sector

The assessment presented here includes the DCI, which utilizes three measures that are either direct or indirect measures of employment and job growth. These are the percent of the prime-age population (25–64) not currently in work, the percent change in the number of jobs over the last five years, and the percent change in the number of business establishments over the last five years. In addition, we used the county unemployment rate. This indicator could be expanded to focus on out-migration due to lack of jobs, long-term underemployment, and long-term decline in business operations due to out-migration or depopulation.

Indicators with Partial or Fragmented Data

Indicator 29: Number of visitor days by activity and recreational land class
This indicator is a good example of one where some data are available over some time periods and apply to only a portion of rangeland counties. Most visitor information is fragmented because it is gathered by public and private entities for different uses and with different methods. In addition, even data gathered systematically may not be useful for larger spatial frameworks. For example, the USDA Forest Service visitor monitoring protocols produce data for national forests and grasslands (English et al., 2002). This cannot be broken down into counties or other smaller units and obviously does not cover private rangelands.

Indicator 40: Community satisfaction

This indicator has a long history in the social sciences. Examples from sociology and economics, respectively, include Brown (1993) and Corrado et al. (2013). An example of gathering this information on a large scale as it relates specifically to natural resources such as rangelands would be Shields et al. (2002). Most research utilizes interview and survey work on groups of communities. The general approach is to capture various dimensions of community satisfaction and to examine interactions among those dimensions. Clearly, any attempt to make this indicator applicable for a national assessment would require sampling and methods that might result in a few communities being examined in rangeland areas of each state. National panel data are available that examine differences in attitudes in counties, but the sample coverage cannot be stretched to account for our spatial needs in this assessment (Parker et al., 2018). Asking for a data set that satisfies our needs at a national spatial scale seems farfetched. However, we can make better use of contextual data to frame issues at the community level and hopefully guide local people to better understand their own communities and the role rangelands play in their community satisfaction.

Indicator 51: Income differentials from migration

The original indicator from Criterion 4 identified "differentials between existing household income in an area" as a possible indicator that long-term local populations have different relationships to natural resources than wealthier in-migrants (Tanaka et al., 2009, p. 24). This may indeed be true and would require a deep examination to see if the data can be used at this spatial scale. There are numerous studies that support the hypothesis embedded in this indicator and could easily support further research into the applicability of those data and methods to rangeland counties (Farrell, 2020; Parker et al., 2018; Smith et al., 2016; Johnson and Lichter, 2020).

Indicator 52: Length of residence (native, immigrant > 5 years, immigrant < 5 years)

This indicator was originally proposed to examine the question of whether the length of residence was related to social cohesion and integration in rangeland communities (Tanaka et al., 2009, p. 25). The ACS can account for the duration of residence of one year. Other US Census sources such as the CPS Annual Economic Supplement and the Survey of Income and Program Participation measure duration of residence with sampling methods (US Census Bureau, 2020; 2020). The ACS duration measure is too short to be an adequate indicator, and the other sources are samples that are not at our spatial level. Nonetheless, this is a relevant research issue. So, once again, it can provide context for local communities to better understand their relationships to the rangelands around them.

Indicator 54: Public beliefs and attitudes about natural resources

This indicator was identified originally because it underlies the notion that long-term sustainability of management decisions has, to some degree, a social acceptability dimension (Tanaka et al., 2009). There are numerous studies of this general question, as well as narrower versions based on beliefs and attitudes about local resources. However, as was noted in the original discussion, it is almost certainly impossible to utilize most studies for a national assessment due to issues of spatial coverage with the samples normally used.

This question is best applied to offer context to an assessment such as this one. As stated earlier, we can make better use of contextual data from these studies to frame issues at the community level and hopefully guide local people to better understand how these public attitudes affect their own communities and the management of rangelands in their area.

Indicators with No Available Data

Indicator 30: Reported threats to quality of recreation experiences

This indicator attempted to understand the actual characteristics of users and to track the impacts of different uses of rangelands for recreation. It was originally envisioned as a "composite index based on the reports of managers of a scientific sample of rangeland recreation settings, stratified by number of discrete units and spatial extent of ownerships" (Tanaka et al., 2009, p. 6). Such data do not exist, and clearly, this index would require a national-scale survey unlikely to be undertaken with either public or private resources.

Indicator 35: Threat or pressure on the integrity of cultural and spiritual resource values

This indicator was intended to identify and track management challenges for rangelands as they pertain to the cultural and spiritual values of rangeland users. Conceptually, this is an important indicator, as it relates public and personal values to the overall value of rangelands. However, no data at a useful scale exist currently (Tanaka et al., 2009, p. 11).

Indicator 42: Presence and tenure of natural resource nongovernmental organizations at the local level

This indicator was intended to measure the capacity of nongovernment organizations (NGOs) for managing local projects involving rangeland resources. There are many NGOs with operational programs that are relevant to rangelands. Even if these organizations were surveyed regarding their programming, it is likely that many counties would have little or no data. To our knowledge, there are no data available.

Economic Data Needs

For the economic data, the Census of Agriculture is collected every five years. The issue we found with this data set is that at the county level, there are a lot of nondisclosures of data. What we really wanted was to find the Census of Agriculture data for producers who owned rangeland or raised beef cattle, sheep, or meat goats at the county level. We were denied this request, as it would lead to mostly nondisclosure of data. Hence, what we used were as much data as we could obtain for the rangeland counties from all producers. Thus, the economic data are from a mix of producers with rangeland, pastureland, crops, and unfarmed land.

Additional data from the Bureau of Labor Statistics and the Bureau of Economic Analysis is collected quarterly and annually, if not more often. These data are useful for some of the economic indicators.

Indicators Used in This Assessment

Indicator 27: Value of forage harvested from rangeland by livestock

We currently used the per acre value of forage on rangeland as reported by the National Agricultural Statistics Service. A better value would be if we knew how many acres were being grazed and for how long to get a total value. This is also only on private land, so getting a value for federal land grazed is also needed. Since federal land grazing is paid on an administratively set fee, it is only a proxy for value. There is a variety of ways to estimate a true value beyond the federal grazing fee, and this should be determined. A recent attempt to value private and public land grazing value at the state level was done by Taylor et al. (2019) and Maher et al. (2020). No similar estimate has been made at the county level, and that would be needed for this sort of national assessment.

Indicator 32: Rate of return on investment for rangeland livestock enterprises

Data are available from the Census of Agriculture to estimate a rate of return on investment at the county level for all agricultural operations. What is not available is a way to determine which of these are rangeland livestock enterprises. A more accurate rate of return could be estimated at the county level if that information were available.

Indicator 33: Area and number of conservation easements

The Census of Agriculture added questions on the acres of conservation easements and the number of farms with conservation easements. While not the same thing as described in Tanaka et al. (2009), these are excellent proxies. However, what is needed is a way to determine how many of these easement acres or farms are rangeland livestock operations.

Indicator 41: Federal transfers by categories (individual, infrastructure, agriculture, etc.)

The Census of Agriculture reports transfers by local, state, and federal governments for agricultural operations. What is not known is what amount of those goes to rangeland operations.

Indicator 44: Employment diversity

The Bureau of Labor Statistics provides data at the county level for fairly high-level industrial classifications. This assessment used a specific index calculation for each rangeland nonurban county. It is not known if this was the correct index to use or if there is a better one. Also, if finer-scale industrial classifications are available at the county level, the index may be more sensitive to changes from one time period to the next.

Indicator 45: Agriculture (ranch/farm) structure

More research is needed to determine if the indicator components used in the assessment are meaningful. The Census of Agriculture provides a wealth of data on the various components at the county level. This index resulted in multiple sub-indicators for this assessment. What is not known is how much of this is applicable to rangeland operations or whether the sub-indicators were the proper ones for such operations.

Indicator 47: Value produced by agriculture and recreation as percent of total

Data are available at the county level to calculate this value from the Bureau of Economic Analysis. What is needed is a way to separate out rangeland-based agriculture and recreation from that total.

Indicator 49: Land tenure, land use, and ownership patterns by size classes

More research is needed to determine if the indicator components used in the assessment are meaningful. The Census of Agriculture provides a wealth of data on the various components at the county level. This index resulted in multiple sub-indicators for this assessment. What is not known is how much of this is applicable to rangeland operations or whether the sub-indicators were the proper ones for such operations.

Indicators with Partial or Fragmented Data

Indicator 28: Value of production of non-livestock products produced from rangeland

As noted in Tanaka et al. (2009), there are no county-level valuations of non-livestock products produced from rangeland. There have been nonmarket valuation studies at specific locations, but none are applicable to every county. Benefit transfer has been used, but that may be of questionable value at the county level. Taylor et al. (2019) and Maher et al. (2020) have developed a methodology to value wildlife and open space on rangelands at the state level, but it cannot be disaggregated to the county level. Hence, a data need is to determine the most important rangeland-based non-livestock products and develop a methodology to consistently estimate the values at the county level.

Indicator 31: Value of investments in rangeland, rangeland improvements, and recreation/tourism infrastructure

The Census of Agriculture provides some estimates of total investments in farm and ranch operations by county. What is not known is what proportion is on rangelands, or for what purpose. A national accounting of such investments on private and public lands is needed for this indicator to be useful. There is no national database on rangeland improvements and recreation or tourism infrastructure investment. The USDA Natural Resources Conservation Service is currently developing a large database on conservation practice investments nationwide. Perhaps when it is fully rolled out, it will prove to be a proxy for such investments on a national scale but likely only on private lands.

Indicator 34: Expenditures (monetary and in-kind) to restoration activities

There is no known source of data at the county, state, or national level for this indicator. There may be proxy values from USDA Natural Resources Conservation Service conservation practices on private land at the county level, but that is not readily available. If it were available, it would be useful to sort out which practices were on rangelands and which qualify as rangeland restoration activities. On federal lands, if a database were available at the county scale on their expenditures on rangelands, that would fill in that gap.

Story Map

A story map is an online application whereby a specific story can be told with the data at hand. We have developed such a story outline here. It can be found at https://www.sustainablerangelands.org/nationalassessment/story/.

The basic intent of that story map is to follow along with what has been presented in this book, though not necessarily in the same framework. In addition, this story map will have links to many more maps and raw data that were used in this assessment. This section lays out that outline.

The Story of Rangeland Sustainability: National Rangeland Assessment

What Is the National Assessment and Why Was It Done?

Selection of Counties for National Assessment
Other Geographical Scales for the Assessment
Ecological, Social, and Economic Indicators
How to Assess Rangeland Sustainability
Rangeland Sustainability – Case Studies
Caveats and Conclusions

Literature Cited

Brown, R., 1993. Rural community satisfaction and attachment in mass consumer society. *Rural Sociology* 58, 387–403. doi: 10.1111/j.1549-0831.1993.tb00501.x

Comer, P., Faber-Langendoen, D., Evans, R., Gawler, S., Josse, C., Kittel, G., Menard, S., Pyne, M., Reid, M., Schulz, K., Snow, K., Teague, J., 2003. *Ecological systems of the United States: a working classification of U.S. terrestrial systems*. NatureServe, Arlington, Virginia.

Corrado, G., Corrado, L., Santoro, E., 2013. On the individual and social determinants of neighborhood satisfaction and attachment. *Regional Studies* 47(4), 544–562. https://doi:10.1080/00343404.2011.587797

Economic Innovation Group, 2018. *From great recession to great reshuffling: charting a decade of change across American communities*. Washington, DC. https://eig.org/wp-content/uploads/2018/10/2018-DCI.pdf (accessed 3 September 2020).

English, D., Kocis, S., Zarnoch, S., Arnold, J., 2002. *Forest service national visitor use monitoring process: research method documentation*. Gen. Tech. Rep. SRS-57. U.S. Department of Agriculture, Forest Service, Southern Research Station, Asheville, NC, p. 14.

[EPA] Environmental Protection Agency, 2020. 303 (d) Impaired waters. Waters Geospatial Downloads. https://www.epa.gov/waterdata/waters-geospatial-data-downloads (accessed 16 September 2020).

[EPA] Environmental Protection Agency, 2021. Water data and tools. https://www.epa.gov/waterdata (accessed 25 January 2021).

Every Texan, 2020. Best uses for the census bureau's current population survey and the American community survey. https://everytexan.org/images/2019_CPS_vs_ACS.pdf (accessed 27 October 2020).

Farrell, J., 2020. *Billionaire wilderness: the ultra-wealthy and the remaking of the American West*. Princeton University Press, Princeton, NJ.

Hafen, K.C., Blasch, K.W., Rea, A., Sando, R., Gessler, P.E., 2020. The influence of climate variability on the accuracy of NHD perennial and nonperennial stream classifications. *Journal of the American Water Resources Association* 56, 903–916.

Here, 2021. Location moves the world forward. https://www.here.com/ (accessed 25 January 2021).

Johnson, K., Lichter, D., 2020. Metropolitan reclassification and the urbanization of rural America. *Demography* 57, 1929–1950, https://doi.org/10.1007/s13524-020-00912-5

Maher, A.T., Maczko, K.A., Taylor, D.T., Tanaka, J.A., 2020. National and state economic values of cattle ranching and farming-based ecosystem services on federal and private lands in the U.S. University of Wyoming B-1367, Sustainable Rangelands Roundtable Publication Number 7.

Microsoft, 2021. Microsoft building footprints – tiles. https://www.arcgis.com/home/item.html?id=f40326b0dea54330ae39584012807126 (accessed 25 January 2021).

NatureServe, 2021. Connecting science with conservation. https://www.natureserve.org/ (accessed 25 January 2021).

Oak Ridge National Laboratory, 2021. LandScan. https://landscan.ornl.gov/ (accessed 25 January 2021).

Parker, K., Horowitz, J., Brown, A., Fry, R., Cohn, D., Igielnik, R., 2018. What unites and divides urban, suburban and rural communities. Pew Research Center. https://www.pewsocialtrends.org/2018/05/22/what-unites-and-divides-urban-suburban-and-rural-communities/ (accessed 3 September, 2020).

Rigge, M., Shi, H., Homer, C., Danielson, P., Granneman, B., 2019. Long-term trajectories of fractional component change in the Northern Great Basin, USA. *Ecosphere* 10(6), e02762. https://doi.org/10.1002/ecs2.2762

Robinson, N.P., Jones, M.O., Moreno, A., Erickson, T.A., Naugle, D.E., Allred B.W., 2019. Rangeland productivity partitioned to sub-pixel plant functional types. *Remote Sensing* 11, 1427. https://doi.org/10.3390/rs11121427.

Rollins, M. 2009. LANDFIRE: a nationally consistent vegetation, wildland fire, and fuel assessment. *International Journal of Wildland Fire* 18(3), 235–249, https://doi:10.1071/WF08088.

Shields, D., Martin, I, Martin, W., Haefele, M., 2002. *Survey results of the American public's values, objectives, beliefs, and attitudes regarding forests and rangelands*. General Technical Report RMRS-GTR-95. USDA Forest Service, Rocky Mountain Research Station, Washington, DC.

Smith, M., Winkler, R., Johnson, K., 2016. *How migration impacts rural America. Brief No. 03-16, Population trends in post-recession rural America, a publication series of the W3001 research project*. Applied Population Lab, University of Wisconsin. https://w3001.apl.wisc.edu/pdfs/b03_16.pdf (accessed 28 September 2020).

Tanaka, J., Torell, L.A., Swanson, L.V., Brunson, M., McCollum, D., Heintz Jr., H.T., 2009. Chapter V: Indicators for maintenance and enhancement of multiple economic and social benefits to current and future generations, in: *Sustainable rangeland roundtable, first approximation report*. University of Wyoming. Laramie, WY. http://www.sustainablerangelands.org/reports/2003Report/Chapter05.pdf (accessed 3 September 2020).

Taylor, D.T., Quintana Ashwell, N.E., Maher, A.T., Maczko, K.A., Tanaka, J.A., 2019. National and state economic values of cattle ranching and farming based ecosystem services in the U.S. University of Wyoming Extension Publication B-1338, Sustainable Rangelands Roundtable Publication Number 5.

US Census Bureau, 2016. Fact sheet: differences between the American Community Survey (ACS) and the Annual Social and Economic Supplement to the Current Population Survey (CPS ASEC). https://www.census.gov/topics/income-poverty/poverty/guidance/data-sources/acs-vs-cps.html (accessed 28 September 2020).

US Census Bureau, 2020. Current population survey. https://www.census.gov/programs-surveys/cps/about.html (accessed 3 September 2020).
U.S. Census Bureau, 2020. Survey of income and program participation (SIPP). https://www.census.gov/programs-surveys/sipp.html (accessed 3 September 2020).
[USGS] U.S. Geological Survey, 2020. FLOwPER user's guide—for collection of FLOw PERmanence field observations. https://www.usgs.gov/center-news/flowper-user-s-guide-collection-flow-permanence-field-observations?qt-news_science_products=1#qt-news_science_products (accessed 25 January 2020).

8

Conclusion and Closing Thoughts

It is important to note that one of the companion processes to the SRR effort, the Roundtable on Sustainable Forests, has completed two national forest assessments (USDA, 2004; Robertson et al., 2011), largely developed by the USDA Forest Service under international agreements that arose out of the Rio Earth Summit and the subsequent Montreal Process. SRR had no such international agreement to provide either incentives or constraints as volunteer participants evolved their work around criteria, indicators, and rangeland assessment. After the SRR First Approximation Report was completed, the SRR pursued individual projects designed to highlight the need and utility of social, ecological, and economic data, whereby knowledge could be gained and enhanced regarding rangeland sustainability. These projects have addressed a broad array of topics including

- development of the conceptual framework for rangeland sustainability (Fox et al., 2009),
- rangeland ecosystem goods and services (Maczko et al., 2011; Maher et al., 2021),
- ranch sustainability assessment framework (Maczko et al., 2012),
- rangelands contributions to food security (Maczko et al., 2013),
- alternative energy development in rangeland systems (Kreuter et al., 2012),
- climate change adaptation and resilience (McCollum et al., 2017),
- socioeconomic surveys (Lind, 2015; Collins, 2019),
- usable science and future directions for rangeland sustainability research (Maczko et al., 2016),
- rangeland soil health (Odom et al., 2018),
- economics of rangeland soil health (Dyer et al., 2021),
- prescribed fire (Wilbur et al., 2021),
- invasive species (Hart and Mealor, 2021), and
- sage-grouse conservation and economics (Quintana Ashwell et al., 2019).

DOI: 10.1201/9781003279747-8

Additional information on these varied projects is available on the SRR website http://sustainablerangelands.org/projects/, as well as in the cited publications. While these efforts have not spawned new national data sets, the research involved has provided information and guidance necessary to facilitate the development of such data sets and offers local and regional templates of how they might be used in assessing rangeland sustainability. Also, when graduate and/or undergraduate students are involved in interdisciplinary research addressing ecological, as well as social and/or economic, aspects of rangeland sustainability, such projects foster engagement and development of knowledge, skills, and abilities for future rangeland scientists, ecologists, economists, and rural sociologists.

As has been noted throughout this book and summarized in Chapter 7, at this time, it is not feasible to gather data on all 64 off SRR's original indicators. In some cases, available data do not match the assumed intent of the indicator. In other cases, data may exist but are not available from the government in the form or at the scale needed due to various laws and regulations regarding privacy. In these latter cases, data can be found at higher levels of aggregation, but they must be interpreted more carefully. Identification of these gaps in necessary information offers a road map for future research and development, highlighting potential opportunities for current and new researchers interested in rangeland sustainability and assessment.

The authors of this book wish to thank all of the participating organizations and individuals who have contributed time and effort to SRR endeavors over the past 20 years. Without their participation and dedication to rangeland sustainability, this assessment would not have been possible. We have made a robust attempt, with limited resources, to conduct this initial national assessment. As has been noted, a more comprehensive approach would require time series of national-scale data to identify and elucidate trends over time. While that is available for some of the indicators, it is not currently accessible for all indicators. Data sets used in this assessment are available on the SRR website at https://www.sustainablerangelands.org/nationalassessment/data/. We hope that subsequent attempts will further develop and complete these time-series data sets to assess rangeland sustainability to characterize the direction and magnitude of change in the indicators.

Lastly, as we have stated repeatedly throughout this book, we do not make conclusions on whether rangelands or rangeland-dependent communities are sustainable. We have attempted to assemble a story map for the assessment; story maps generally are designed to create inspiring, immersive stories by combining text, interactive maps, and other multimedia content. The assessment story map is available via the SRR website at https://www.sustainablerangelands.org/nationalassessment/story/.

We have provided tools for individuals, communities, regions, and/or stakeholder groups at other geographical scales to evaluate the data and determine if they are moving in the desired direction. Individual or group values will affect how data are interpreted. For example, if we view the indicator on conservation easements, those acreage numbers may increase in the future. If you believe that restricting land use for a conservation purpose helps society, more acres would be viewed as good. On the other hand, if you are a land developer, such legal protections may hinder your business and be viewed less favorably. Some indicators will be interpreted fairly consistently (impaired waters is normally a "bad") and others will have disparate views. In all cases, the desired degree and direction of change will be subject to discussion.

We believe that the guiding principles (Table 1.2) that provided a solid foundation for the SRR hold true today and into the future. Widely accepted criteria and indicators of rangeland sustainability and supporting data to populate the indicators will permit people to have rational and informed discussions about changes needed to conserve, maintain, and enhance rangelands and the communities dependent upon them. Better data facilitate informed discussions for enhanced and defensible decision-making.

Literature Cited

Collins, A., 2019. Demographics, motivations, and experiences of producers participating in the environmental quality incentives program and the conservation stewardship program. MS Thesis. Laramie, WY: University of Wyoming.

Dyer, H., Maher, A.T., Ritten, J.P., Tanaka, J.A., Maczko, K.A., 2021. Ranch profitability of improving soil health on rangelands. *Rangeland Ecology and Management* 77(1), 66–74.

Fox, W.E., McCollum, D.W., Mitchell, J.E., Swanson, L.E., Kreuter, U.P., Tanaka, J.A., Evans, G.R., Heintz, H.T., Breckenridge, R.P., Geissler, P.H., 2009. An integrated social, economic, and ecologic conceptual (ISEEC) framework for considering rangeland sustainability. *Society & Natural Resources* 22(7), 593–606.

Hart, M., Mealor, B.A., 2021. Effects of *Ventenata dubia* removal on rangelands of northeast Wyoming. *Invasive Plant Science and Management* 14(3), 156–163.

Kreuter, U.P., Fox, W.E., Tanaka, J.A., Maczko, K.A., McCollum, D.W., Mitchell, J.E., Duke, C.S., Hidinger, L., 2012. Framework for comparing ecosystem impacts of developing unconventional energy resources on western US rangelands. *Rangeland Ecology and Management* 65, 433–443.

Lind, B., 2015. Social and economic characteristics of public land ranchers nationally and in Wyoming. MS Thesis. Laramie, WY: University of Wyoming.

Maczko, K., Hidinger, L., Tanaka, J., Ellis, C., 2016. A workshop on future directions of usable science for rangeland sustainability. *Rangelands* 38(2), 53–63.

Maczko, K.A., Tanaka, J.A., Ritten, J., Fox, W.E., Morgan, J.A., Chhetri, N., Hidinger, L.A., Reeves, M.C., Mitchell, J.E., 2013. Assessing sustainable rangelands contributions to food security, in: Peterson, J.W. (Ed.), *Proc. 5th National Conference on Grazing Lands*, December 6–9 2012. Orlando, FL. Grazing Lands Conservation Initiative, Bryan, TX.

Maczko, K.A. Tanaka, J.A., Smith, M., Garretson-Weibel, C. Hamilton, S.F., Mitchell, J.E., Fults, G., Stanley, C., Loper, D., Bryant, L.D., Brite, J.K., Jr. 2012. Ranch business planning and resource monitoring for rangeland sustainability. *Rangelands* 34(1), 11–16.

Maczko, K., Tanaka, J.A., Breckenridge, R., Hidinger, L., Heintz, H.T., Fox, W.E., Kreuter, U.P., Duke, C.S., Mitchell, J.E., McCollum, D.W. 2011. Rangeland ecosystem goods and services: Values and evaluation of opportunities for ranchers and land managers. *Rangelands* 33(5), 30–36.

Maher, A.T., Quintana Ashwell, N., Maczko, K.A., Taylor, D.T., Tanaka, J.A., Reeves M.C., 2021. An economic valuation of federal and private grazing land ecosystem services supported by beef cattle ranching in the United States. *Translational Animal Science* 5(3), 1–15.

McCollum, D.W., Tanaka, J.A., Morgan, J.A., Mitchell, J.E., Fox, W.E., Maczko, K.A., Hidinger, L., Duke, C.S., Kreuter, U.P., 2017. Climate change effects on rangelands and rangeland management: Affirming the need for monitoring. *Ecosystem Health and Sustainability* 3(3), 1–13.

Odom, L., Maczko, K., Derner, J., Dell, C., McCulley, R. Kucera, J., Franzluebbers, A., Knaebel, D., Manter, D., 2018. Assessing and managing for soil health on rangelands and pasture lands: A workshop white paper. Foundation for Food and Agriculture Research. https://foundationfar.org/wp-content/uploads/2018/06/Rangeland-Soil-Health-White-Paper-6.6.18.pdf.

Quintana Ashwell, N., Maher, A.T., Tanaka, J.A., Ritten, J.P., Maczko, K.A., Dyer, H., Kirkpatrick, H., Roberts, K., Hilken, T., 2019. Ranch economics of prescribed grazing on rangelands: The case of representative ranches in Northern Montana. *Western Economics Forum, Western Agricultural Economics Association* 17(1), 73–85.

Robertson, G., Gualke, P., McWilliams, R., LaPlante, S., Guldin, R. (eds.), 2011. *National report on sustainable forests—2010. FS-979*. USDA Forest Service, Washington, DC, p. 212.

USDA Forest Service, 2004. National report on sustainable forests – 2003. FS-766. Washington, DC: USDA Forest Service, February 2004, p. 139.

Wilbur, R., Stanley, C., Maczko, K.A., Scasta, J.D. 2021. Perceptions of NRCS assistance with prescribed fires on U.S. private lands: A regionally stratified case study. *Fire* 4(3), 47–59.

Appendix: List of Abbreviations

CD	Congressional Districts
DCI	Distressed Communities Index
EIG	Economic Innovation Group
EMAP	Environmental Monitoring and Assessment Program
ESA	Ecological Society of America
EVC	Existing Vegetation Cover
EVH	Existing Vegetation Height
EVT	Existing Vegetation Type
FGDC	Federal Geographic Data Committee
FIA	Forest Inventory and Analysis
ISEEC	Integrated Social, Economic, and Ecological Conceptual [framework]
KE	Küchler Ecoregions
MEA	Millennium Ecosystem Assessment
MLRA	Major Land Resource Area
NASS	National Agricultural Statistics Service
NPS	National Park Service
NRC	National Research Council
NRI	National Resource Inventory
NVC	National Vegetation Classification
RPA	Resource Planning Act
RSF	Roundtable on Sustainable Forests
SDI	Simpson's Diversity Index
SRR	Sustainable Rangelands Roundtable
USDA	United States Department of Agriculture
USDI	United States Department of the Interior
VDep	Vegetation Departure

Index

Pages in *italics* refer to figures and **bold** refer to tables.